# Discrete Relaxation Techniques

THE INTERNATIONAL SERIES OF MONOGRAPHS ON COMPUTER SCIENCE



1. J. Vitter and W. C. Chen
   *Design and Analysis of Coalesced Hashing*
2. H. Reichel
   *Initial Computability*
3. J. O'Rourke
   *Art Gallery Theorems and Algorithms*
4. C. Colbourn
   *The Combinatorics of Network Reliability*
5. T. C. Henderson
   *Discrete Relaxation Techniques*

# Discrete Relaxation Techniques

THOMAS C. HENDERSON

*University of Utah*

New York Oxford

OXFORD UNIVERSITY PRESS

1990

Oxford University Press

Oxford New York Toronto
Delhi Bombay Calcutta Madras Karachi
Petaling Jaya Singapore Hong Kong Tokyo
Nairobi Dar es Salaam Cape Town
Melbourne Auckland

and associated companies in
Berlin Ibadan



Published by Oxford University Press, Inc.,
200 Madison Avenue, New York, New York 10016



Library of Congress Cataloging-in-Publication Data
Henderson, Thomas C.
Discrete relaxation techniques.
(The International series of monographs on computer science ; 5)
Bibliography: p. Includes index.
1. Integrated circuits—Very large scale integration—Mathematical models.
2. Relaxation methods (Mathematics) 3. Artificial intelligence—Mathematical models.
4. Computer vision—Mathematical models. I. Title. II. Series.
TK7874.H399 1989 621.395 88-32964 ISBN 0-19-504894-6

9 8 7 6 5 4 3 2 1

Printed in the United States of America

*To Jeanne d'Arc*

# Preface

Relaxation techniques have been used extensively in solving certain numerical analysis problems for quite some time. In this book we describe recent developments in the application of relaxation techniques to finite domains. Thus, the origin of the term "discrete relaxation." These techniques have found wide application in artificial intelligence and computer vision.

The major goals of this book are to provide the formal foundations of these techniques, to provide some examples of their application, and finally to summarize the state-of-the-art algorithms, both sequential and parallel, and special-purpose architectures for performing discrete relaxation.

*Salt Lake City*
*December 1988*

T.C.H.

# Acknowledgment

In writing this monograph on discrete relaxation, I have many colleagues and associates to thank for their support. In particular, I would like to thank my students Ashok Samal and Jun Gu with whom I developed many of the algorithms and architectures found in Chapters 8 and 9. They have provided a steady stream of ideas and dialogue. Thanks also to Wei Wang for some key ideas on the architectural formulation of discrete relaxation.

In addition, I would like to thank Roger Mohr for getting me interested in the complexity of arc consistency algorithms (OK, and path consistency, too). The time spent in Nancy, France was both enlightening and entertaining. Perhaps we will have to explore some more of that territory in the future.

Finally, a blanket thanks to colleagues at the University of Utah with whom I have had many interesting discussions on this and related topics.

# Contents

# Discrete Relaxation Techniques

# Chapter 1

# Introduction

Classical relaxation techniques were introduced by R.V. Southwell[79] over fifty years ago as a practical aid to the engineer in computing the stresses in braced frameworks. As investigation proceeded, it was shown that the method had a much wider range of application than the original domain for which it was designed. Likewise, the symbolic (as opposed to numeric) version of relaxation introduced by Waltz[88] as a method for labeling edges in line drawings has been shown to be a powerful technique of much wider significance. Our main goal is to present a unified view of relaxation, both as a numerical and as a symbolic technique. In doing so, we will try to demonstrate the range and power of this approach. The examples are chosen mostly from the field of computer vision since the majority of the work on symbolic relaxation has been done by workers in that field.

Constraints play a key role in relaxation techniques. In classical relaxation, the local constraints are introduced by expanding at each point the partial differential equation which gives the global constraint. Likewise, symbolic relaxation can be seen as a method for imposing constraints. Such constraints can be derived from the physical world or can be the result of purely symbolic considerations. That constraints can be usefully applied is witnessed by several remarkable systems based on relaxation. Sutherland's Sketchpad[83] system provides a graphical interface which recognizes and enforces geometric constraints to be satisfied. The ThingLab environment developed by Borning[4] is another example of such a system, only in this case the system provides a more complete graphical simulation capability to the user. Finally, Gosling[24] has implemented a system based on more general algebraic constraints. All of these systems rely on relaxation techniques to maintain the user-defined constraints. Other constraint propagation systems have been implemented for VLSI design and have achieved notable success[80].

There are two major types of relaxation:

- Classical relaxation: The numerical technique by which a solution is produced for a set of linear simultaneous equations
- Symbolic relaxation: The labeling technique by which a labeling is determined for a set of objects

Both forms of relaxation are useful in computer vision. Classical relaxation is an important technique in so-called "low-level" processing; e.g., the computation of stereo disparity, optical flow, etc. As we shall see in greater detail later on, classical relaxation is appropriate for the solution of elliptic partial differential equations, and many properties of interest in visual data can be so characterized. Symbolic relaxation can be quite effective in solving the labeling problem; i.e., given a set of possible labels and a set of objects extracted from the visual data, use the known constraints between labels and objects to produce a consistent labeling of the objects. In exploring the two forms of relaxation, we shall see that they have a common basis, and that they share some of the same drawbacks.

The relaxation techniques that interest us are characterized by the following properties:

- A single global constraint which the solution must satisfy is transformed into a set of many local constraints
- A simple operation is applied uniformly at all unknowns to produce a solution.

Later chapters develop the meaning of these two properties in much greater detail. Suffice it to say for the moment that in classical relaxation the global constraint corresponds to an elliptic partial differential equation, whereas in the earliest symbolic relaxation schemes, the global constraints took the form of physically possible edge combinations. These lead to a set of local arithmetic relations, and a local set of edge junction possibilities, respectively.

As to the second property, the guiding principle for the application of the numerical relaxation process is to drive to zero the residual error of an estimated solution. The form of the updating rule is identical at all unknowns. Likewise, a consistent labeling is achieved in symbolic relaxation by means of a definition of consistency which must be satisfied at each object. Iterative matrix methods for achieving these two operations are developed in Chapters 2 through 4.

Relaxation techniques, both classical and symbolic, have found wide application in computer vision. In general, computer vision research is motivated by several important principles:

- Invert the image formation process
- Reduce the amount of data
- Convert physical signals to symbolic structures

Much of the early work done in pattern recognition and image processing concerned direct manipulation of digital images and digital data. Such an approach utilizes only a partial amount of the total information potentially recoverable. In order to understand the real world, it is then necessary to model the physical objects and processes involved in the image formation process. In this way, it can be known under what conditions the image formation process can be inverted to recover the three-dimensional reality which produced a particular image. This approach gives rise to the diverse "shape from" techniques: shape from shading, shape from texture, etc. Many of these techniques are based upon physical models which are expressed as partial differential equations. Thus, classical relaxation has been used quite extensively as the numerical tool to solve these problems.

Although data reduction is a guiding principle in most disciplines, it is a very strong requirement in computer vision, especially when a real-time processing constraint is imposed. Data reduction can occur in many places in the image analysis, including at the signal processing level of data acquisition and encoding, at the primitive extraction level, or at the high-level symbolic processing level. Another problem faced in common by practitioners of classical relaxation and computer vision is that of noise and ambiguity in the data. Symbolic relaxation has been found particularly useful in allowing the application of higher-level or anthropomorphic notions of structure to be brought to bear on the primitive extraction problem; e.g., noise elimination in pixel intensities, edge parameter amelioration, etc. These applications can be considered as low-level symbolic relaxation.

Finally, once physical signals or their digital versions are available, they are converted to symbolic structures. The analysis of these structures and their relations constitutes the major research area of computer vision and draws heavily on artificial intelligence techniques. It is in this application perhaps that symbolic relaxation has been applied with the most success. Relaxation and many relaxation operators have been investigated as a means of speeding up search.

Although the domain of application in which we are interested is computer vision, no detailed discussion of that topic will be given. Our emphasis will be on the relaxation techniques themselves, and each application to a specific problem will be described in as self-contained a manner as possible.

Thus far, we have only distinguished between classical and symbolic relaxation. There is, however, a significant subdivision of symbolic relaxation into:

- Discrete relaxation: A discrete set of labels is assigned to each object, and the updating rule deletes elements from those sets
- Stochastic relaxation: Each object has an associated vector which assigns to each label the likelihood of that label for the object. The updating rule then modifies the likelihood vectors according to the local context of the object.

These two forms of symbolic relaxation actually represent two radically different approaches to solving the labeling problem. The introduction of likelihoods poses several significant hurdles, both in theory and in practice. The updating rule determines the convergence properties of the relaxation process, and therefore must be chosen carefully. Furthermore, many intuitively appropriate updating rules are difficult to characterize through formal analysis. Also, the a priori likelihoods are difficult to assign in a reasonably safe way. In order to solve most of these problems, stochastic relaxation is formulated as a constrained optimization problem. Stochastic relaxation is not covered in this book, however, several people have analyzed this problem, including Barrow, Faugeras and Berthod, Hummel and Zucker, or Rosenfeld[2,13,43,67].

In this brief introduction, we have attempted to give an overview of the terrain to be covered and the related nature of the material. We hope to synthesize the results from many sources and provide a useful bibliography of the fast-growing symbolic relaxation and constraint satisfaction literature. We believe that these techniques have much to offer in the solution of difficult computer vision problems, and when used appropriately can significantly reduce the computation involved.

# Chapter 2

# Classical Relaxation

## 2.1 Southwell's Heuristic Method

The numerical method known as *relaxation* was introduced in the early 1940's by R.V.L. Southwell [78,79] as a means for solving (by hand) large systems of linear algebraic simultaneous equations. The method was intended for direct practical application to real world problems, and although in theory based on a single idea, the ways of applying the method were as diverse as the individuals who used the method and the number of different problems they encountered. The many variations on the essential operation of the method meant that no two users were guaranteed to solve a given problem in exactly the same way, but the answer was correct to whatever degree of required accuracy. The method was also applicable to finite difference approximations to partial differential equations, and that application was probably even more important than the use of relaxation in solving general simultaneous linear equations, especially in light of the fact that the special form of the difference equations permitted a much more regular and efficient application of the relaxation method. Another important feature of the relaxation method was that one could work with whole numbers and if arithmetic errors occurred, it was not necessary to recalculate previous work, one could merely continue on and eliminate the error in the normal application of the method. Southwell's method was not intended to produce an exact analytic solution, but rather a solution within the desired tolerance of error for the given application.

### 2.1.1 Basic Relaxation

Consider the case of a set of $n$ simultaneous linear equations in $n$ unknowns:

$$\mathbf{A}\mathbf{x} = \mathbf{b} \tag{2.1}$$

which we rewrite as:

$$\begin{gathered} a_{10} + a_{11}x_1 + \cdots + a_{1n}x_n = 0 \\ \vdots \\ a_{n0} + a_{n1}x_1 + \cdots + a_{nn}x_n = 0 \end{gathered}$$

where $-\mathbf{b} = [a_{10}a_{20}\ldots a_{n0}]^t$. For any fixed assignment of values to the $x_i, i = 1,\ldots n$, we define the $i$th *residual* to be:

$$R_i = a_{i0} + a_{i1}x_1 + \cdots + a_{in}x_n$$

The classical relaxation process simply defines a way by which the set of linear simultaneous equations can be solved by (1) choosing an initial set of values for the unknowns and then (2) applying a sequence of operations which changes the values of the unknowns and drives the residuals toward zero. Obviously, the set of values for which the residuals are zero is a solution to the problem. In general, though, the application of the method ceases when each $R_i$ is less than some specified tolerance.

In order to characterize the relaxation process, we must first describe explicitly the net effect on each residual of a unit change on a given unknown. Consider the $i$th unknown, $x_i$. A change in value of 1 at $x_i$ results in a change of $a_{11}, a_{21}, \ldots, a_{n1}$ in residuals $R_1, R_2, \ldots, R_n$, respectively.

This can be summarized in the basic unit operations table as follows:

| | $\delta R_1$ | $\delta R_2$ | $\ldots$ | $\delta R_i$ | $\ldots$ | $\delta R_n$ |
|---|---|---|---|---|---|---|
| $\delta x_1 = 1$ | $a_{11}$ | $a_{21}$ | | $a_{i1}$ | | $a_{n1}$ |
| $\delta x_2 = 1$ | $a_{12}$ | $a_{22}$ | | $a_{i2}$ | | $a_{n2}$ |
| $\vdots$ | $\vdots$ | $\vdots$ | | $\vdots$ | | $\vdots$ |
| $\delta x_n = 1$ | $a_{1n}$ | $a_{2n}$ | | $a_{in}$ | | $a_{nn}$ |

Thus, in order to drive a given residual to zero, we must simply choose a basic unit operation and apply it the appropriate number of times. The calculations in a relaxation application then consist of a record of the sequence of basic unit operations chosen and the consequent changes in the residuals.

The goal to be achieved in applying the relaxation process is to change all of the residuals toward zero (also called the liquidation of residual). Moreover, one basic rule can be used to arrive at the goal:

- The basic rule of relaxation: At each step choose a basic unit operation and apply it (or its negative) enough times (consecutively) to reduce the largest residual as close to zero as possible.

As an example of the process, consider the pair of equations:

$$10 \cdot x_1 + 2 \cdot x_2 = 422$$
$$1 \cdot x_1 + 10 \cdot x_2 = 297$$

Then the basic unit operations table for this pair of equations is:

| | $\delta R_1$ | $\delta R_2$ |
|---|---|---|
| $\delta x_1 = 1$ | 10 | 2 |
| $\delta x_2 = 1$ | 1 | 10 |

To apply the relaxation process, we must first choose a pair of initial values for $x_1$ and $x_2$. Assuming that we have no evidence as to the solution values, we will choose $x_1 = x_2 = 0$ as the initial value. The sequence of operations can be recorded as follows:

| Operation | | $R_1$ | $R_2$ | |
|---|---|---|---|---|
| $x_1, x_2 =$ | 0 | $-422$ | $-297$ | (step 0) |
| $\delta x_1 =$ | 43 | 8 | $-254$ | (step 1) |
| $\delta x_2 =$ | 26 | 60 | 6 | (step 2) |
| $\delta x_1 =$ | -6 | 0 | 0 | (step 3) |

At step 0 we have:

$$R_1 = 10 \cdot 0 + 2 \cdot 0 - 422 = -422$$
$$R_2 = 1 \cdot 0 + 10 \cdot 0 - 297 = -297$$

Thus, we use $43 \cdot \delta x_1$ to bring $R_1$ as close to zero as possible:

$$R_1 = 10 \cdot 43 + 2 \cdot 0 - 422 \;=\; 8$$
$$R_2 = 1 \cdot 43 + 10 \cdot 0 - 297 \;=\; -254$$

Next, we reduce $R_2$ by $26 \cdot \delta x_2$:

$$R_1 = 10 \cdot 43 + 2 \cdot 26 - 422 \;=\; 60$$
$$R_2 = 1 \cdot 43 + 10 \cdot 26 - 297 \;=\; 6$$

and finally, driving $R_1$ to zero with $-6 \cdot \delta x_1$:

$$R_1 = 10 \cdot 37 + 2 \cdot 26 - 422 = 0$$
$$R_2 = 1 \cdot 37 + 10 \cdot 26 - 297 = 0$$

Column 1 records the operation chosen at each step, while columns 2 and 3 record the values of the residuals, $R_1$ and $R_2$, respectively, after the application of the operator. Each step of the process is recorded as a separate line which can be read left to right and gives the operation chosen and the result. The first line simply gives the initial values selected for the unknowns and the residuals associated with those initial values.

To find the solution values, take the initial value for each unknown and add to it the scaling factors for operations associated with that unknown:

$$x_i = x_{i0} + \sum_{j=1}^{steps} \delta x_{ij}$$

where $\delta x_{ij}$ is the operation on $x_i$ at the $j$th step.

For our example we have:

$$\text{Solution for } x_1 = 0 + 43 + 0 - 6 = 37$$
$$\text{Solution for } x_2 = 0 + 0 + 26 + 0 = 26$$

It is always a good idea to check the result by substituting back in the equations. If the solution is not correct, then it is not necessary to look for the error in the already worked sequence, it suffices to compute the correct residuals and continue with the relaxation process.

If the above described process were all there were to relaxation it would never have achieved such widespread use since the work involved is still quite tedious and for many equations and unknowns would still be a horrific task. However, many extensions and modified uses of the basic relaxation process were shown to permit a much more rapid conclusion of the process. We will not examine these in detail since the purpose of this chapter is simply to introduce classical relaxation; however, we will attempt in the later sections of this book to explain how some of these extensions motivate analogous extensions to symbolic relaxation algorithms.

*Overrelaxation* is a technique by which one goes past zero in order to overcompensate for subsequent changes in the other residuals.

*Block relaxation* defines the simultaneous application of one or more basic unit operations (but the same number of times). Thus, given our previous example, we could define a *basic unit* block operations table:

| Operation | $\delta R_1$ | $\delta R_2$ |
|---|---|---|
| $\delta x_1 = 1$ | 10 | 2 |
| $\delta x_2 = 1$ | 1 | 10 |
| $\delta x_1 = \delta x_2 = 1$ | 11 | 12 |

from which multiples are readily obtained:

$$\delta x_1 = \delta x_2 = 3 \qquad 33 \qquad 36$$

There are three block operations defined by the table, one each for the unknowns separately and one for the pair.

*Group relaxation* is a mechanism by which a sequence of distinct operations can be used as one operation. Such an operation is essentially the same as a block operation except that the number of applications of distinct basic unit operations may differ. (Block relaxation is a special case of group relaxation.)

These are just some of the possible variations on the basic relaxation process, and many more are reported in the literature[74,78,79]. These others are, however, mostly irrelevant to our purposes and we now turn to a more central problem: the role of global constraints in generating local constraints.

### 2.1.2 Relaxation Applied to Ordinary Differential Equations

The most important application of the relaxation technique is to the solution of differential equations. We restrict our attention to partial differential equations of no more than two dimensions. The nature of the typical problem to be solved is illustrated by Figure 2.1. The relaxation method of obtaining the solution to such a problem is to determine values of the desired function $W$ at a finite collection of points equally spaced throughout the surface under consideration.

Consider for a moment the one-dimensional problem of determining $W(x)$ from $x = a$ to $x = b$ and given the end (or boundary) conditions of $W(a)$ and $W(b)$. The solution produced by the relaxation process is a set of values at $x = a + kh$, $k = 1$ to $n - 1$, where $h$ is the distance between successive points; that is, $b = a + nh$.

In order to apply the relaxation method to solve a differential equation, we must approximate the derivatives by finite difference equations. Usually, only the first

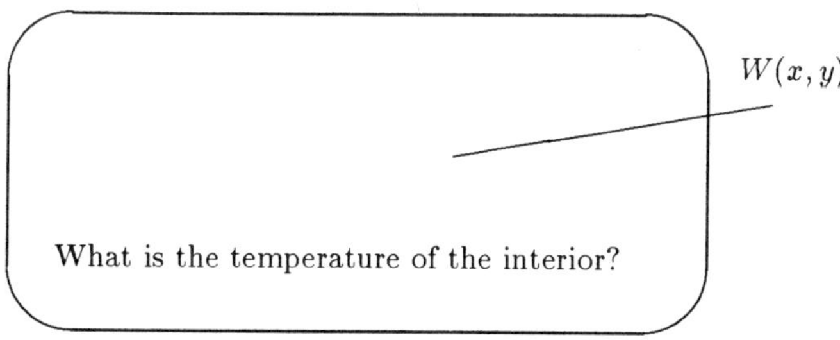


Figure 2.1: The problem is to determine the temperature $W(x, y)$ in the interior of the disk.

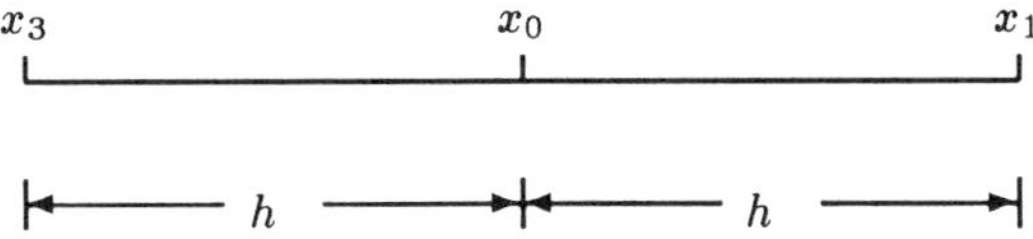


Figure 2.2: A point and its two neighbors.

and second derivatives need be approximated. Following Allen[1], we will derive these approximations in terms of a point and its two neighbors (see Figure 2.2). Whenever we refer to a general point $x_0$, its neighbors will be called $x_3$ and $x_1$. (We use $x_3$ to be consistent with the notation for the two-dimensional case given below.) Then,

$$f(x) = f(x_0) + f'(x_0)(x - x_0) + \frac{f''(x_0)(x - x_0)^2}{2!} + \cdots + \frac{f^n(x_0)(x - x_0)^n}{n!}$$

If we evaluate this series for $f$ at $x_3$ and $x_1$, we get:

$$f(x_1) = f(x_0) + hf'(x_0) + \frac{h^2 f''(x_0)}{2} + \frac{h^3 f'''(x_0)}{6} + \cdots \tag{2.2}$$

$$f(x_3) = f(x_0) - hf'(x_0) + \frac{h^2 f''(x_0)}{2} - \frac{h^3 f'''(x_0)}{6} + \cdots \tag{2.3}$$

Adding (2.2) to (2.3), we obtain:

$$f(x_1) + f(x_3) = 2f(x_0) + h^2 f''(x_0) + \text{ fourth-order terms and up}$$

Therefore, we can use

$$h^2 f''(x_0) = f(x_1) + f(x_3) - 2f(x_0) \tag{2.4}$$

as an approximation to the second derivative. To get a difference equation for the first derivative, we subtract (2.3) from (2.2) and neglect higher-order terms:

$$f(x_1) - f(x_3) = 2hf'(x_0)$$

from which we get:

$$hf'(x_0) = \frac{W_1 - W_3}{2}$$

where

$$W_1 = W(x_1) \quad \text{and} \quad W_3 = W(x_3)$$

There are many properties (e.g., the transverse deflection function) which are characterized by an equation of the form:

$$f''(x` + W(x) = 0$$

where $W$ is some known function of $x$. When this is combined with (2.4) the result is:

$$f(x_1) + f(x_3) - 2f(x_0) + h^2 W(x_0) = 0$$

In general, the method is applied to produce solutions at $n - 1$ unknowns across an interval. Note that the form of the finite difference equation is identical at each point. This leads to the observation that during the relaxation process the computation of the residual is accomplished by using a *relaxation pattern* at the unknown under consideration; i.e., the pattern

$$\begin{bmatrix} 1 & -2 & 1 \end{bmatrix}$$

is applied at an unknown. This means that the residual is formed as:

$$R = f(x_3) - 2f(x_0) + f(x_1)$$

Although the use of finite difference equations is straightforward, there is another way of looking at it which will play an important role in symbolic relaxation. Namely, the differential equation provides a global characterization of some property and from that one derives a set of local relations which must hold. That is, one goes from a single equation which holds everywhere to a set of equations, each one of which holds only at one point. We will see that this notion of deriving many local relations (or constraints) from one global relation is the key to much of the symbolic relaxation work.

The final topic from classical relaxation which is of interest to us is the use of Laplace's equation in two dimensions:

$$f_x^{''} + f_y^{''} = 0$$

The corresponding relaxation patterns are:

$$f_x^{''} = \begin{bmatrix} 1 & -2 & 1 \end{bmatrix} \qquad f_y^{''} = \begin{bmatrix} 1 \\ -2 \\ 1 \end{bmatrix}$$

Thus, their sum yields the two-dimensional relaxation pattern:

$$\begin{bmatrix} & 1 & \\ 1 & -4 & 1 \\ & 1 & \end{bmatrix}$$

This is the standard Laplacian operator used in image processing. Given that Laplace's equation must hold over some 2-D region, then the relaxation process must find values at some finite set of points within the region of interest. Such a set of points is called a grid, and it is usually square or triangular (also hexagonal, but not often). When these grid points are connected to nearest neighbors, the result is called a *relaxation net*; we develop a somewhat different terminology for the subsequent generalization of this idea.

## 2.2 Matrix Iterative Techniques

With the advent of computers, there arose another approach to the solution of elliptic partial differential equations. Namely, a theory of matrix iterative analysis was developed and provided a mechanism for the "efficient solution, by means of iteration, of matrix equations arising from discrete approximations to partial differential equations" (see Varga[87]). Varga's book is an excellent survey, and we follow his development of the subject.

Whereas Southwell's methods were developed for use by humans and inherently required much imagination and skill in their application, the use of computers demanded the formulation of an algorithm which could be programmed on a digital computer. Thus, a major difference between the two approaches is that Southwell's methods were heuristic, while the matrix techniques were algorithmic.

As we have seen, a matter of great concern to any practitioner of the relaxation method is the convergence of the method. The expanded version of the (2.1) is:

$$\mathbf{A} = \begin{bmatrix} a_{11} & a_{12} & \dots & a_{1n} \\ a_{21} & a_{22} & \dots & a_{2n} \\ \vdots & \vdots & \vdots & \vdots \\ a_{n1} & a_{n2} & \dots & a_{nn} \end{bmatrix} \quad \mathbf{x} = \begin{bmatrix} x_1 \\ \vdots \\ x_n \end{bmatrix} \quad -\mathbf{b} = \begin{bmatrix} a_{10} \\ \vdots \\ a_{n0} \end{bmatrix}$$

The values of $\mathbf{b}$ are calculated from the boundary conditions. A very important property of the matrix derived from self-adjoint second-order elliptic partial differential equations is that it can be written in the form:

$$\mathbf{A} = \mathbf{I} - \mathbf{B} \tag{2.5}$$

where $\mathbf{B}$ has only nonnegative real number entries and is symmetric, and, moreover, has its eigenvalues in modulus less than unity, and $\mathbf{I}$ is the identity matrix.

Combining (2.1) and (2.5), we have:

$$\mathbf{x} = \mathbf{B}\mathbf{x} + \mathbf{b} \tag{2.6}$$

The *point Jacobi* iterative method is based on (2.6) in that a sequence of iterates is defined:

$$\mathbf{x}^{(k+1)} = \mathbf{B}\mathbf{x}^{(k)} + \mathbf{b}$$

Once such a sequence is defined, its convergence properties can be studied. In order to answer this, it is necessary to determine whether $\lim_{k\to\infty} x_i^k$ exists for each $i$, etc. It can be shown that in order for the sequence to converge, then the following must hold:

$$\lim_{k\to\infty} \mathbf{B}^k = 0$$

To really tie down the meaning of convergence for sequence of matrixes, we need some definitions.

**Definition 2.1** *If* $\mathbf{A}$ *is an* $n \times n$ *complex matrix, and the eigenvalues of* $\mathbf{A}$ *are given by* $\lambda_i, i = 1, n$*, then the* spectral radius *of* $\mathbf{A}$, $\rho(\mathbf{A})$ *is given by:*

$$\rho(\mathbf{A}) = \max_{1\le i\le n} |\lambda_i|$$

**Definition 2.2** *If* $\mathbf{A}$ *is an* $n \times n$ *complex matrix, then* $\mathbf{A}$ *is* convergent *if the sequence of matrixes* $\mathbf{A}, \mathbf{A}^2, \mathbf{A}^3, \dots$ *converges to the null matrix* $\mathbf{0}$*; otherwise* $\mathbf{A}$ *is* divergent.

We have then the following theorem:

**Theorem 2.1** *If* $\mathbf{A}$ *is an* $n \times n$ *complex matrix, then* $\mathbf{A}$ *is convergent iff* $\rho(\mathbf{A}) < 1$.

Thus, we now have a much clearer picture of the meaning of convergence. Much of the remaining theory of matrix iterative methods deals with determining the spectral radius and with the behavior of convergent sequences.

In this chapter, we do not intend to develop the full theory for matrix iterative analysis (for a full account, see Varga[87]); instead we will now examine some of the iterative methods. In later chapters we will demonstrate their relation to symbolic relaxation processes.

Let us assume, then, that the problem is to solve (2.1) with the matrix $\mathbf{A}$ nonsingular and the diagonal entries nonzero. $\mathbf{A}$ can be rewritten as:

$$\mathbf{A} = \mathbf{D} - \mathbf{L} - \mathbf{U} \tag{2.7}$$

where $\mathbf{D}$ is the diagonal of $\mathbf{A}$, $\mathbf{L}$ is the (negative) strictly lower part of $\mathbf{A}$, and $\mathbf{U}$ is the (negative) strictly upper part of $\mathbf{A}$. Combining (2.1) and (2.7) we have:

$$\mathbf{D}\mathbf{x} = (\mathbf{L} + \mathbf{U})\mathbf{x} + \mathbf{b}$$

Thus, we can express each unknown as:

$$a_{ii}x_i^{(k+1)} = -\sum_{j \neq i} a_{ij}x_j^{(k)} + b_i$$

or finally as:

$$\mathbf{x}^{(k+1)} = \mathbf{D}^{-1}(\mathbf{L} + \mathbf{U})\mathbf{x}^{(k)} + \mathbf{D}^{-1}\mathbf{b}$$

This formulation is called the *point Jacobi* or the *point total-step iterative method.*

An alternative formulation is:

$$a_{ii}x_i^{(k+1)} = -\sum_{j=1}^{i-1} a_{ij}x_j^{(k+1)} - \sum_{j=i+1}^{n} a_{ij}x_j^{(k)} + b_i$$

or

$$(\mathbf{D} - \mathbf{L})\mathbf{x}^{(k+1)} = \mathbf{U}\mathbf{x}^{(k)} + \mathbf{b}$$
$$\Rightarrow \mathbf{x}^{(k+1)} = (\mathbf{D} - \mathbf{L})^{-1}\mathbf{U}\mathbf{x}^{(k)} + (\mathbf{D} - \mathbf{L})^{-1}\mathbf{b}$$

This formulation is called the *point Gauss-Seidel* or *point single-step iterative method.*

These two methods, the point Jacobi and the point Gauss-Seidel, are the basis for the symbolic relaxation techniques. Although there are other methods, e.g., overrelaxation, and subtle questions of convergence of the methods and the rate of convergence, we have now seen the essentials of classical relaxation as we shall need them.

## 2.3 Applications of Classical Relaxation in Computer Vision

Optical flow is the retinal field of apparent intensity velocities produced by the relative motion of the objects and the viewer in space. There is much psychological and neurophysiological evidence that optical flow plays an important role in the perception of space in biological vision systems. Horn and Schunck[40] have described a method for computing optical flow based on constraints from which two equations can be derived in terms of the unknown spatial velocity components in the $x$ and

$y$ directions. Relaxation techniques are used to solve these equations due to the inherently unreliable approximations to various derivatives used in the equations. Our development follows that of Horn and Schunck.

First, we assume that a given patch of surface on an object in the visual field maintains a constant intensity throughout its apparent motion across the retina. Let $f(x, y, t)$ be the retinal intensity function which is dependent on two spatial dimensions, $x$ and $y$, and on time, $t$. Then the above assumption leads to:

$$f(x, y, t) = f(x + \delta x, y + \delta y, t + \delta t)$$

Expanding the right-hand side in a Taylor expansion gives:

$$f(x, y, t) = f(x, y, t) + \delta x f_x + \delta y f_y + \delta t f_t + \text{ higher-order terms}$$

where

$$f_x = \frac{\delta f''}{\delta x} \quad f_y = \frac{\delta f''}{\delta y} \quad f_t = \frac{\delta f''}{\delta t}$$

from which we get

$$0 = \frac{\delta x}{\delta t} f_x + \frac{\delta y}{\delta t} f_y + f_t + \text{term of order } \delta t$$

In the limit then, we have that

$$0 = \frac{\delta x}{\delta t} f_x + \frac{\delta y}{\delta t} f_y + f_t$$

or

$$0 = f_x u + f_y v + f_t$$

where

$$u = \frac{dx}{dt} \text{ and } v = \frac{dy}{dt}$$

Thus, we have derived one equation in two unknowns in terms of the first assumption or constraint. In order to delimit a solution, we must impose at least one more constraint.

A second reasonable assumption is that neighboring points on the retinal field will most likely have similar velocities. This is known as the smoothness constraint. We will use the magnitude of the gradient as a measure of smoothness. Thus, we seek to minimize:

$$\left(\frac{\delta u}{\delta x}\right)^2 + \left(\frac{\delta u}{\delta y}\right)^2 \text{ and } \left(\frac{\delta v}{\delta x}\right)^2 + \left(\frac{\delta v}{\delta y}\right)^2$$

In order to actually solve the problem, we have to approximate various quantities, including $f_x$, $f_y$, and $f_t$:

$$f_x = \frac{f(x+1, y+j, t+k) - f(x, y+j, t+k)}{4} \quad j, k = 0, 1$$

$$f_y = \frac{f(x+i, y+1, t+k) - f(x+i, y, t+k)}{4} \quad i, k = 0, 1$$

$$f_t = \frac{f(x+i, y+j, t+1) - f(x+i, y+j, t)}{4} \quad i, j = 0, 1$$

The approximations to $\nabla^2 u$ and $\nabla^2 v$ are given by:

$$\nabla^2 u \approx \overline{u} - 4u_{i,j,k}$$
$$\nabla^2 v \approx \overline{v} - 4v_{i,j,k}$$

where

$$\overline{u} = u_{i-1,j,k} + u_{i,j+1,k} + u_{i+1,j,k} + u_{i,j,k}$$
$$\overline{v} = v_{i-1,j,k} + v_{i,j+1,k} + v_{i+1,j,k} + v_{i,j,k}$$

We can then define an error function $E$ which must be minimized:

$$E^2 = \sum \sum (\alpha^2 E_2 + E_1) dx dy$$

where $E_i$ is the error in terms of the $i$th constraint. We can then use the calculus of variations to obtain:

$$f_x^2 u + f_x f_y v = \alpha^2 \nabla^2 u - f_x f_t$$
$$f_y^2 v + f_x f_y u = \alpha^2 \nabla^2 v - f_y f_t$$

Substituting for $\nabla^2 u$ and $\nabla^2 v$, we get:

$$(\alpha^2 + f_x^2)u + f_x f_y v = \alpha^2 \overline{u} - f_x f_t$$
$$(\alpha^2 + f_y^2)v + f_x f_y u = \alpha^2 \overline{v} - f_y f_t$$

Solving for $u$ and $v$, we get:

$$(\alpha^2 + f_x^2 + f_y^2)u = (\alpha^2 + f_y^2)\overline{u} - f_x f_y \overline{v} - f_x f_t$$
$$(\alpha^2 + f_x^2 + f_y^2)v = (\alpha^2 + f_x^2)\overline{v} - f_x f_y \overline{u} - f_y f_t$$

If we allow $\alpha^2$ to tend to zero and apply the method of Lagrange multipliers to the problem of minimizing $E_2^2$ while maintaining $E_1 = 0$, we get:

$$f_y \nabla^2 u = f_x \nabla^2 v \quad \text{and} \quad f_x u + f_y v + f_t = 0$$

which implies that

$$(f_x^2 + f_y^2)(u - \overline{u}) = -f_x(f_x \overline{u} + f_y \overline{v} + f_t)$$
$$(f_x^2 + f_y^2)(v - \overline{v}) = -f_y(f_x \overline{u} + f_y \overline{v} + f_t)$$

We can then use the standard Gauss-Seidel method to compute new velocity estimates $[u^{(n+1)}, v^{(n+1)}]$ by:

$$u^{(n+1)} = \overline{u} - \frac{f_x(f_x \overline{u}^n + f_y \overline{v}^n + f_t)}{f_x^2 + f_y^2}$$
$$v^{(n+1)} = \overline{v} - \frac{f_y(f_x \overline{u}^n + f_y \overline{v}^n + f_t)}{f_x^2 + f_y^2}$$

This approach is difficult to apply in practice and is computationally expensive.

# Chapter 3

# The Labeling Problem

## 3.1 Things and Names

The world as we perceive it is made up of objects, and any symbolic approach to analyzing the world requires that objects be distinguished and that some symbol, or name, be associated with each object. The localization of distinct objects is called the segmentation problem while the assignment of a name to an object is called the labeling problem.

In the problems we will consider, the lowest level objects are usually obtained directly from some kind of sensor, but they can also arise in an abstract problem domain. For example, a camera may provide light intensities at various locations in an image. Each of these locations constitutes a picture element (or pixel), and these intensities provide the basis for all subsequent analysis. Objects that are not defined in terms of other objects are called primitives, or primitive elements. Other examples of primitives are edges, pieces of shape, and surface patches. Primitives also arise in abstract problem domains. A Boolean variable in a well-formed formula is an example of such a primitive.

There are two important classes of names given to objects:

1. Physical labels: The desired label of an object is just a measurement of or an approximation to some physical quantity. In this case, the range of possible labels is just the range of the measured value to whatever quantization is possible.

2. Symbolic labels: The labels have no underlying physical significance (or only indirectly), but rather are the result of an arbitrary set of naming conventions or describe the functionality of the object (e.g., truth values for variables in Boolean expressions).

Although physical labels can be treated as symbolic labels, there are good reasons, both historical and practical, for treating the two separately. We will explore some of these reasons in the next few chapters.

Physical labels arise in many contexts. For example, if the primitive is intensity in an image, then the value of the intensity may be the label of interest. Suppose that intensity is quantized to eight gray levels, then there are 256 possible labels for any particular pixel. As we will show later, certain image processing operations can be viewed as methods for relabeling an initial set of intensities in order to

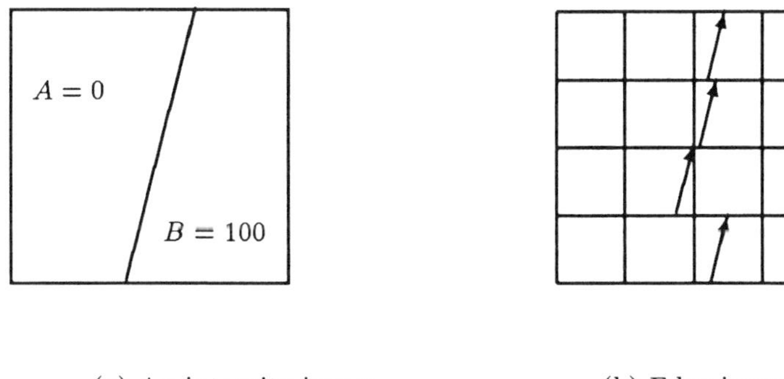


(a) An intensity image (b) Edge image

Figure 3.1: Intensity and edge image.

satisfy constraints between the pixels. Observe that classical relaxation can be considered in this light as a labeling problem in which the physical labels represent the heat, stress, etc. of the particular problem. Thus, we see that the labeling problem includes the standard set of problems usually solved by classical relaxation techniques.

As another example of a physical label, consider edge element primitives located in a two-dimensional image as primitives. Each such edge element can be very coarsely modeled as a directed vector through the pixel. Figure 3.1a gives an intensity image and Figure 3.1b shows the corresponding edge image. For each pixel in the edge image then, there may or may not be an edge element present. If there is an edge element present, then it is usually quantized into some small set of possible values, e.g., $\{ \rightarrow, \nearrow, \uparrow, \nwarrow, \leftarrow, \swarrow, \downarrow, \searrow \}$, where the darker side of the intensity edge is understood to be to the left of the directed vector. If we are trying to decide what is the appropriate edge vector for each pixel, then we are solving a physical labeling problem. Typically, edge results are cleaned up by averaging an edge with the nearby responses; this also can be viewed as a labeling problem. Another version of this problem is to assign labels from the set convex edge, concave edge, occluded edge, occluding edge, shadow edge. Again, this is a physical labeling problem. However, sometimes the label to be assigned to the edge element describes what part of a complete shape the edge is; e.g., part of the wing of an airplane shape. In this case, the label to be assigned is not a result of the physical nature of the world, but rather an arbitrary name which is logically, or symbolically, related to other edge elements to form some global pattern. We would like to distinguish between these two kinds of labeling.

As just described, there is a second kind of labeling problem, called the symbolic labeling problem. In general, a physical labeling problem is distinguished from a symbolic labeling problem in the manner in which one derives the constraints between the objects to be labeled and their possible labels. Physical labels derive from the physics of the situation, whereas symbolic labels are usually the result of convention or are related to the anthropomorphic view of an object. For example, a spoon has a "handle." Furthermore, in completely symbolic settings (e.g., given a well-formed formula whose tautological status is to be determined), the constraints may follow directly from a set of well-established rules. That is, the notion of "satisfiability" for a well-formed formula depends on the interpretation of the logical connectives and the laws of Boolean algebra.

One final example is the cryptarithmetic problem. Given a set of symbols substituted for the digits in an arithmetic expression, assign a unique digit to each symbol (in this case, letters of the alphabet) such that the resulting arithmetic expression is correct. For example:

$$\begin{array}{r} \mathbf{OAR} \\ +\ \mathbf{BOW} \\ \hline \mathbf{WIN} \end{array} \qquad \text{has the solution} \qquad \begin{array}{r} 146 \\ +213 \\ \hline 359 \end{array}$$

This is a very general kind of labeling problem, and as can be seen, it is defined in terms of three things: (1) the objects to be labeled, (2) the possible labels for each object, and (3) the constraints placed on labeling several objects simultaneously. Although it is possible to use a brute force approach to the problem and to try all possible combinations of letter-digit pairings to find the solution, the amount of search can be reduced by the use of constraints. The extent to which such constraints can be helpful depends on the problem to be solved and the constraints inherent in that problem. Constraints on the solution of a problem generally make it easier to find a solution for the problem. In fact, sometimes no explicit solution is even possible without enough constraints on the solution. We now discuss the role played by constraints in such a formulation of the labeling problem.

## 3.2 Constraints

One convenient way to think about constraints between objects is to view them as asserted relations between labels and objects. Most people have a fairly well developed understanding of the nature and meaning of constraints in the physical world. We are all aware that there are constraints on the amount of weight we can lift, on the speed we can run, on how far we can jump, etc. These are physical constraints on our bodies. Moreover, there are other constraints which perhaps go unnoticed, but which we exploit in understanding the world around us. For example, nearer objects usually occlude more distant objects; straight (or nearly straight) lines usually continue smoothly. At a different level of description, cars are usually near roads or houses, and airplane engines are usually parallel to the airplane's fuselage. These are examples of the kinds of constraints which arise in our descriptions of the world.

Just as human beings use constraints in understanding the world and dealing with it, so it is also possible to exploit such constraints to solve labeling problems. Obviously, the kind of problem to be solved will determine the kinds of constraints and their relations to the objects and labels. That is, it may be necessary to give a great deal of study and thought to discover what the appropriate objects and labels are. Moreover, the problem domain will most probably also determine the nature of the constraints and what they mean. However, our goal is to find a general methodology for applying such constraints to the solution of labeling problems. In addition, we would like to provide some insight into reasonable ways of obtaining objects and labels to make them more amenable to the application of known constraints for the solution of a particular problem.

Now let's go back for a second look at the cryptarithmetic problem. Suppose that we want to solve the following problem:

$$\begin{array}{r} \mathbf{A} \\ +\mathbf{B} \\ \hline \mathbf{BC} \end{array}$$

Clearly the objects are the letters of the alphabet and the labels are the ten digits $\{0, 1, \ldots, 9\}$. The constraints should include the uniqueness postulate, i.e.,

- A 1-1 map exists between letters and digits.

as well as the constraints derived from the rules of arithmetic. A small sample of these is:

- If two letters generate a carry, then neither letter corresponds to zero (assuming no carry into their sum).

- If a letter added to itself generates a carry, then the corresponding digit is in the set $\{5, 6, \ldots, 9\}$ (assuming no carry into their sum).

and so on.

In solving such problems as the cryptarithmetic problem, people generally rely on such constraints to arrive at a solution (or set of possible solutions), and rarely have to rely on brute force enumeration tactics. Thus, it is clear that the determination of such constraints and the development of a framework in which they can be suitably exploited constitute desirable goals for any problem solving system.

Consider for a moment the physical constraint of smooth continuation for edges in the real world. Using the set of edge labels given earlier for a two-dimensional edge image, we can make the notion of "smooth continuation" much more precise. The eight edge vectors are:

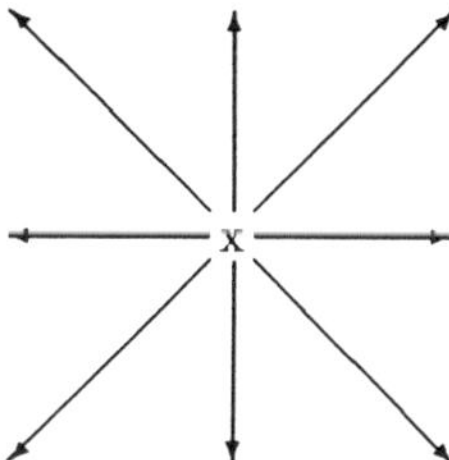


We can define smooth continuation for an edge at pixel $p(i, j)$ by considering the immediate neighbors of $p(i, j)$. For example, we might require that the two neighboring pixels along the direction of the edge at $p(i, j)$ have the same direction. That is, one of the following must hold:

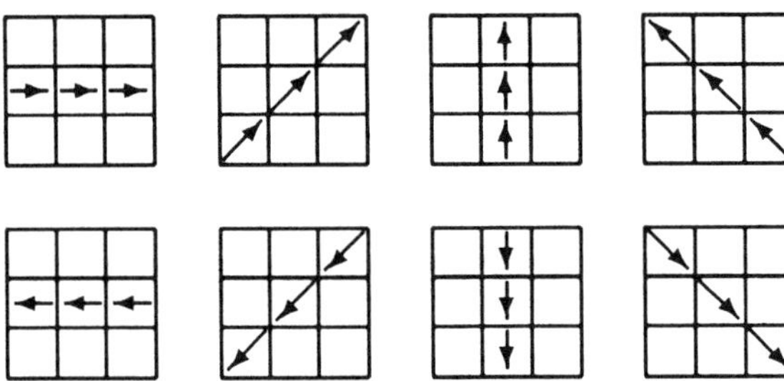

A label corresponding to one of the orientations of $0^\circ, 45^\circ, \ldots, 315^\circ$ at pixel $(i, j)$ is said to have smooth continuation if the neighboring pixels in the directions of the edge ahead and behind have labels corresponding to an orientation within $\pm 45^\circ$ of it. We can use this notion to perform certain kinds of image processing operations on the image, e.g., by forcing smooth continuation at each pixel. In general, it is good to explicitly define the opposite notion, which here does not really correspond to "not smooth continuation," but rather to "discontinuity." Such a concept can be defined much the same as for "smooth continuation," but with edge vectors at adjacent pixels pointing in opposite directions. This would be quite useful, for example, in locating corners of squares on a checkerboard.

Relations may be defined on the objects themselves, independently of the labels. For example, "long" might be a useful unary relation. Given an object, $a$, the unary relations may be decided on the basis of relative measurements between the whole set of objects or in terms of some absolute threshold. For example, an object might be considered long only if it is longer than two-thirds of the other objects, or simply if its length exceeds some specified value. We must not overlook the fact that some things, such as edge vector orientation, can be viewed as features of an object, a unary relation on an object, or as a label, depending on the problem under consideration.

Relations may also be defined between objects. Thus, two objects may be "neighbors" or one may be "left_of" the other or "above" the other. Some frequently used geometric relations are "parallel" and "perpendicular." Generally constraints will be defined by asserting the relations that groups of labels must satisfy. Then, the labeling problem is solved by assigning labels to objects and checking if the constraints will help avoid the investigation of a large number of the possible labelings. Put another way, constraint satisfaction techniques can be used to prune search trees.

In general, it is possible to define $n$-ary relations between objects, and we will present a couple of approaches to providing such a capability. However, such an approach usually is computationally prohibitive, and thus a lot of effort has gone into finding other ways to provide the same power. We discuss these in later chapters.

## 3.3 Consistency

A very powerful method for marshalling the use of evidence is that of determining consistency. From the previous discussion we have seen that the solution to the labeling problem has four steps:

1. Determine the appropriate objects and labels for the problem.

2. Determine the constraints between objects and labels.

3. Make an initial set of hypotheses, i.e., assign a set of possible labels to each object.

4. For each object, eliminate hypothesized labels which violate some constraint.

Thus, it is step 4 which, upon termination, guarantees consistency among the remaining possible labels. Note that there may not exist a consistent unique labeling, but that taken as a whole, each label is consistent with respect to at least one of the other possible labels. A formal version of this method will be given in a later chapter; for the moment, we merely want to discuss the general nature of this approach and to examine some of its ramifications and limitations.

The general method behind the discrete relaxation process is the following:

- Hypothesize every possible label for each object, and then eliminate any label which violates some imposed constraint.

Usually, the constraints are abstracted from the problem and must always hold. Sometimes, absolute knowledge must be incorporated, too, and the notion of consistency of imposed, known data must be reconciled with the constraints of the system. Problems can arise, for example, if an incorrect label is associated with an object, then a whole multitude of other incorrect labels may be maintained as hypotheses even though they would otherwise be eliminated.

Note that it is not guaranteed that a resulting labeling will provide a unique label for each object, but only that each remaining label is consistent with some hypothesized label at every other object. In the symbolic labeling problem, this can cause major difficulties since, in general, each object starts with the full set of labels hypothesized, and no label will be eliminated because every other label exists for the other objects, and thus every constraint is automatically satisfied. For the physical labeling problem, this may not pose problems in practice because the initial label sets for the different objects are very small and the initial constraints are directly related to the relations between the objects rather than the relations between the labels.

## 3.4 Convergence of Discrete Relaxation

Convergence is easily proven for most formulations of discrete relaxation. This is due to the fact that a hypothesized label is discarded at every step, and it must eventually be the case that there are no more labels to discard; i.e., the empty set of labels is achieved on some object. Thus, discrete relaxation is very different from stochastic relaxation, where the convergence properties of certain operators are very difficult to determine. However, even if the resulting set of labels is consistent, there is no guarantee that any consistent assignment of unique labels to objects exists. Thus, some kind of tree search is usually performed after the discrete relaxation

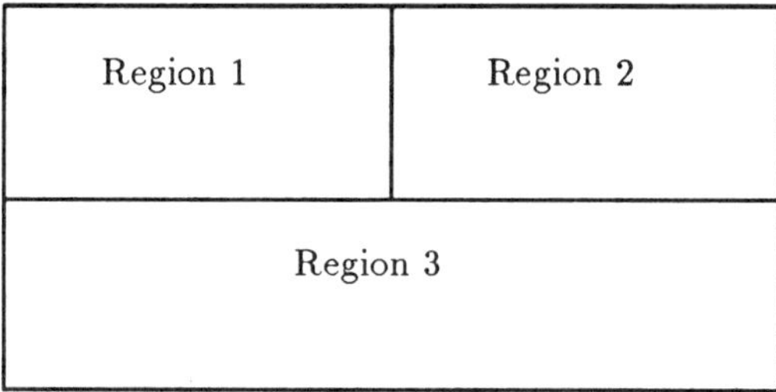


Figure 3.2: Three neighboring regions to be colored.

process has finished or combined with relaxation. For example, suppose that three neighboring regions (the objects) are to be colored red or blue (the labels) as shown in Figure 3.2. Furthermore, the coloring is to be done subject to the following constraints:

1. Each region is to be colored only one color
2. Two neighboring regions must be different colors.

Initially, we may hypothesize that any given region can be colored either red or blue (see Figure 3.3). Now, if we apply step 4 to determine the consistency of the labels, we see that no label can be eliminated: Each red label is consistent with the blue label in the other regions, and likewise each blue label is consistent with the red label in the other regions. However, as can be readily seen, there is no globally consistent labeling with only two colors, since each region neighbors the other two regions. This can be determined either by tree search or by using higher-order relations.

Several versions of discrete relaxation have been proposed, but they are all equivalent. The method should be applied only after it has been determined that the discrete relaxation approach has some advantages over other methods. One good indication that discrete relaxation may be appropriate is the strength of the constraints. The stronger they are, the more appropriate the method. Another good indication is the degree of "localness" of the constraints. If the constraints must be

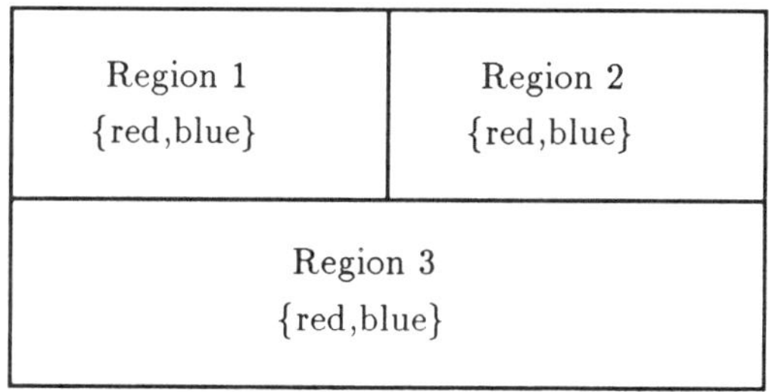


Figure 3.3: Possible colors for each region.

defined in terms of arbitrary numbers of objects, then the amount of computation will most likely be too great. Finally, and this is related to the localness of the constraints, if the computation can be performed well by a set of parallel processors working on local data, then the method is most likely appropriate.

One issue which has not been raised yet, but will be discussed in much greater detail later, is the automatic generation of constraints. In certain symbolic labeling problems, it is possible to specify a set of explicit constraints, which are obvious to a human, and then have any implicit constraints be determined automatically; e.g., by transitivity (see Chapter 6). This can be a very powerful approach when combined with the symbolic labeling algorithm. In many computer vision applications, it is also necessary to take into account the noise and ambiguity of the data. Thus, more general discrete relaxation schemes will also be considered.

# Chapter 4

# The Formulation of Discrete Relaxation

Classical relaxation techniques have been studied for many years, but only recently have researchers realized the potential for such methods in solving machine vision problems. The earliest applications of symbolic relaxation methods involved a "discrete" relaxation that we show to be a restriction of the general relaxation process to systems of Boolean inequalities which take values over the two element set $\{0, 1\}$. (This chapter is a modified version of Reference 35.)

## 4.1 Introduction

The classical relaxation techniques introduced by R.V. Southwell as a practical aid to the engineer in computing stresses have been described in Chapter 2. Such techniques have proven useful in solving a wide range of problems and constitute a standard numerical method. In a similar way, the symbolic (as opposed to numeric) versions of relaxation introduced by Guzman[28] and Waltz[88] as methods for labeling edges in line drawings are known to be examples of a more general technique[67]. In this chapter we present a unified view of relaxation, both as a numerical and as a symbolic technique. For a review of the numerous applications of symbolic relaxation processes in computer vision, see Davis and Rosenfeld[8].

The classical relaxation problem typically involves a spatial arrangement of grid elements (here a square grid when $n = k^2$) such as that shown in Figure 4.1. The goal is to determine a simultaneous assignment of values to the $u_i$'s such that some constraint function, e.g., Laplace's equation, is satisfied. Such a function is

$$\begin{array}{cccc} u_1 & u_2 & \cdots & u_k \\ u_{k+1} & u_{k+2} & \cdots & u_{2k} \\ & \vdots & & \\ u_{n-k+1} & u_{n-k+2} & \cdots & u_n \end{array}$$

Figure 4.1: Typical grid layout for classical relaxation.

transformed from a global expression, for example, $\frac{\delta^2 x}{\delta x^2} + \frac{\delta^2 y}{\delta y^2} = 0$, to a set of local finite difference equations, $u_i - \frac{(u_{i1}+u_{i2}+u_{i3}+u_{i4})}{4} = 0$, where element $u_i$ is expressed in terms of its neighbors defined by the $ik$ subscripts. In the plane, these neighbors correspond to the nearest grid points which lie "above," "below," "right," and "left" of $u_i$, when those exist. For boundary points, certain of these neighbors may not exist, and therefore, boundary points may have fewer neighbors. Thus, each of the finite difference equations can be expressed as one row of a coefficient matrix which represents the relation of one particular unknown. That is, for $u_i$:

$$m_{i,i}u_i + \sum_{j \neq i} m_{i,j}u_j = v_i \tag{4.1}$$

If the value of $u_i$ is known in an absolute sense, i.e., not just in terms of its neighbors, then we use $u_i = v_i$ instead of Equation 4.1, where $v_i$ is the known value of $u_i$. Thus, for $n$ unknowns, an $n \times n$ matrix of coefficients is generated, and the $i$th row has nonzero elements at the positions of unknowns related to the $i$th unknown. The set of simultaneous linear difference equations can be written:

$$\mathbf{Mu} = \mathbf{v} \tag{4.2}$$

where $\mathbf{M}$ is the matrix of coefficients of the equations, $\mathbf{u}$ is the column vector of unknowns, and $\mathbf{v}$ is the column vector of known constants. The matrix $\mathbf{M}$ is assumed nonsingular and the diagonal entries nonzero. In the classical case, various methods exist for solving such a system, usually involving an iterative scheme when inversion of $\mathbf{M}$ is too expensive or impractical.

## 4.2 Boolean Formulation of Discrete Relaxation

Suppose now that a real number solution at each unknown is no longer sought, but rather that the solution to be found involves the assignment of a set of labels at each unknown such that some constraint relation among the labels is satisfied by neighboring unknowns. Whereas the unknowns in Equation 4.2 take on real number values, the unknowns in a labeling problem take on a Boolean vector value with each element in the vector corresponding to a possible label. Boolean vector operations are denoted by $\neg$, $\times$, $^t$, $*$, $+$, and $\cdot$ which represent complementation, vector multiplication, transpose, Boolean "and", Boolean "or", and Boolean vector dot product, respectively. Following Henderson[35], let:

1. $U = \{u_1, \ldots, u_n\}$ be the set of unknowns
2. $\Lambda = \{\lambda_1, \ldots, \lambda_m\}$ be the set of possible labels
3. $\Lambda_i = (l_1, \ldots, l_m)^t$ be the column vector describing the set of labels (i.e., zero or one) possible for $u_i$, where $l_j = 1$ if $\lambda_j$ is compatible with $u_i$; 0 otherwise
4. $\Lambda_{ij}$ be an $n \times n$ compatibility matrix for label pairs, where $\Lambda_{ij}(\lambda, \lambda') = 1$ if $\lambda$ at $u_i$ is compatible with $\lambda'$ at $u_j$; 0 otherwise
5. $\Lambda_{ij}(k)$ denotes the $k$th row of $\Lambda_{ij}$.

For example, consider the case of three regions to be colored red, green, or blue with the constraints:

1. Region 1 must be red
2. Region 3 must be green
3. No two regions may be colored the same color.

Thus, $u_i$ = Region $i$ (for $i = 1, 3$) and:

$$U = \{u_1, u_2, u_3\}$$
$$\Lambda = \{\lambda_1, \lambda_2, \lambda_3\}$$

where $\lambda_1$ is red, $\lambda_2$ is blue, and $\lambda_3$ is green. Since Region 1 must be red, we have:

$$\Lambda_1 = \begin{bmatrix} 1 \\ 0 \\ 0 \end{bmatrix}$$

and since Region 3 must be green:

$$\Lambda_3 = \begin{bmatrix} 0 \\ 0 \\ 1 \end{bmatrix}$$

Finally, since there is no restriction on Region 2's color, we have all possibilities:

$$\Lambda_2 = \begin{bmatrix} 1 \\ 1 \\ 1 \end{bmatrix}$$

Since only similar colors are incompatible, we have:

$$\Lambda_{ij} = \begin{bmatrix} 0 & 1 & 1 \\ 1 & 0 & 1 \\ 1 & 1 & 0 \end{bmatrix} \text{ for different objects}$$

$$\Lambda_{ii} = \begin{bmatrix} 1 & 0 & 0 \\ 0 & 1 & 0 \\ 0 & 0 & 1 \end{bmatrix} \text{ for the same object}$$

Since $\Lambda_i$ and $\Lambda_j$ are known and may be less than the full labeling, we can recalculate $\Lambda_{ij}$ as:

$$\Lambda_{ij} := (\Lambda_i^t \times \Lambda_j) * \Lambda_{ij}$$

where the symbol ":=" means the assignment operator. We have then that:

$$\begin{aligned} \Lambda_{11} &= \left(\begin{bmatrix} 1 & 0 & 0 \end{bmatrix}^t \times \begin{bmatrix} 1 & 0 & 0 \end{bmatrix}\right) * \Lambda_{11} \\ &= \begin{bmatrix} 1 & 0 & 0 \\ 0 & 0 & 0 \\ 0 & 0 & 0 \end{bmatrix} * \begin{bmatrix} 1 & 0 & 0 \\ 0 & 1 & 0 \\ 0 & 0 & 1 \end{bmatrix} = \begin{bmatrix} 1 & 0 & 0 \\ 0 & 0 & 0 \\ 0 & 0 & 0 \end{bmatrix} \end{aligned}$$

$$\begin{aligned} \Lambda_{12} &= \left(\begin{bmatrix} 1 & 0 & 0 \end{bmatrix}^t \times \begin{bmatrix} 1 & 1 & 1 \end{bmatrix}\right) * \Lambda_{12} \\ &= \begin{bmatrix} 1 & 1 & 1 \\ 0 & 0 & 0 \\ 0 & 0 & 0 \end{bmatrix} * \begin{bmatrix} 0 & 1 & 1 \\ 1 & 0 & 1 \\ 1 & 1 & 0 \end{bmatrix} = \begin{bmatrix} 0 & 1 & 1 \\ 0 & 0 & 0 \\ 0 & 0 & 0 \end{bmatrix} \end{aligned}$$

$$\begin{aligned}
\Lambda_{13} &= \left(\begin{bmatrix} 1 & 0 & 0 \end{bmatrix}^t \times \begin{bmatrix} 0 & 0 & 1 \end{bmatrix}\right) * \Lambda_{13} \\
&= \begin{bmatrix} 0 & 0 & 1 \\ 0 & 0 & 0 \\ 0 & 0 & 0 \end{bmatrix} * \begin{bmatrix} 0 & 1 & 1 \\ 1 & 0 & 1 \\ 1 & 1 & 0 \end{bmatrix} = \begin{bmatrix} 0 & 0 & 1 \\ 0 & 0 & 0 \\ 0 & 0 & 0 \end{bmatrix}
\end{aligned}$$

$$\begin{aligned}
\Lambda_{21} &= \left(\begin{bmatrix} 1 & 1 & 1 \end{bmatrix}^t \times \begin{bmatrix} 1 & 0 & 0 \end{bmatrix}\right) * \Lambda_{21} \\
&= \begin{bmatrix} 1 & 0 & 0 \\ 1 & 0 & 0 \\ 1 & 0 & 0 \end{bmatrix} * \begin{bmatrix} 0 & 1 & 1 \\ 1 & 0 & 1 \\ 1 & 1 & 0 \end{bmatrix} = \begin{bmatrix} 0 & 0 & 0 \\ 1 & 0 & 0 \\ 1 & 0 & 0 \end{bmatrix}
\end{aligned}$$

$$\begin{aligned}
\Lambda_{22} &= \left(\begin{bmatrix} 1 & 1 & 1 \end{bmatrix}^t \times \begin{bmatrix} 1 & 1 & 1 \end{bmatrix}\right) * \Lambda_{22} \\
&= \begin{bmatrix} 1 & 1 & 1 \\ 1 & 1 & 1 \\ 1 & 1 & 1 \end{bmatrix} * \begin{bmatrix} 1 & 0 & 0 \\ 0 & 1 & 0 \\ 0 & 0 & 1 \end{bmatrix} = \begin{bmatrix} 1 & 0 & 0 \\ 0 & 1 & 0 \\ 0 & 0 & 1 \end{bmatrix}
\end{aligned}$$

$$\begin{aligned}
\Lambda_{23} &= \left(\begin{bmatrix} 1 & 1 & 1 \end{bmatrix}^t \times \begin{bmatrix} 0 & 0 & 1 \end{bmatrix}\right) * \Lambda_{23} \\
&= \begin{bmatrix} 0 & 0 & 1 \\ 0 & 0 & 1 \\ 0 & 0 & 1 \end{bmatrix} * \begin{bmatrix} 0 & 1 & 1 \\ 1 & 0 & 1 \\ 1 & 1 & 0 \end{bmatrix} = \begin{bmatrix} 0 & 0 & 1 \\ 0 & 0 & 1 \\ 0 & 0 & 0 \end{bmatrix}
\end{aligned}$$

$$\begin{aligned}
\Lambda_{31} &= \left(\begin{bmatrix} 0 & 0 & 1 \end{bmatrix}^t \times \begin{bmatrix} 1 & 0 & 0 \end{bmatrix}\right) * \Lambda_{31} \\
&= \begin{bmatrix} 0 & 0 & 0 \\ 0 & 0 & 0 \\ 1 & 0 & 0 \end{bmatrix} * \begin{bmatrix} 0 & 1 & 1 \\ 1 & 0 & 1 \\ 1 & 1 & 0 \end{bmatrix} = \begin{bmatrix} 0 & 0 & 0 \\ 0 & 0 & 0 \\ 1 & 0 & 0 \end{bmatrix}
\end{aligned}$$

$$\begin{aligned}
\Lambda_{32} &= \left(\begin{bmatrix} 0 & 0 & 1 \end{bmatrix}^t \times \begin{bmatrix} 1 & 1 & 1 \end{bmatrix}\right) * \Lambda_{32} \\
&= \begin{bmatrix} 0 & 0 & 0 \\ 0 & 0 & 0 \\ 1 & 1 & 1 \end{bmatrix} * \begin{bmatrix} 0 & 1 & 1 \\ 1 & 0 & 1 \\ 1 & 1 & 0 \end{bmatrix} = \begin{bmatrix} 0 & 0 & 0 \\ 0 & 0 & 0 \\ 1 & 1 & 0 \end{bmatrix}
\end{aligned}$$

$$\begin{aligned}
\Lambda_{33} &= \left(\begin{bmatrix} 0 & 0 & 1 \end{bmatrix}^t \times \begin{bmatrix} 0 & 0 & 1 \end{bmatrix}\right) * \Lambda_{33} \\
&= \begin{bmatrix} 0 & 0 & 0 \\ 0 & 0 & 0 \\ 0 & 0 & 1 \end{bmatrix} * \begin{bmatrix} 1 & 0 & 0 \\ 0 & 1 & 0 \\ 0 & 0 & 1 \end{bmatrix} = \begin{bmatrix} 0 & 0 & 0 \\ 0 & 0 & 0 \\ 0 & 0 & 1 \end{bmatrix}
\end{aligned}$$

The $(p, q)$ entry of $\Lambda_{ij}$ tells if $\lambda_p$ at object $i$ is compatible with $\lambda_q$ at object $j$. For example, $\Lambda_{11}$ reveals that only $\lambda_1$ is compatible with $\lambda_1$ at object 1; i.e., that Region 1 must be colored red.

A *labeling* is a vector $\mathbf{L} = (\mathbf{L}_1, \ldots, \mathbf{L}_n)^t$, where $\mathbf{L}_i = (l_{i1}, \ldots, l_{im})^t$ in $\Lambda_i$ is a Boolean vector with $l_{ij} = 1$ if label $\lambda_j$ is a possible label for object $u_i$; 0 otherwise. A labeling is *consistent* if for every $i$ and $k$:

$$l_{ik} \leq \prod_{j=1}^{n} \left[ \sum_{p=1}^{m} (l_{ik} * l_{jp} * \Lambda_{ij}(k, p)) \right] \tag{4.3}$$

Once a formal definition of local consistency such as Equation 4.3 has been given, it is easy to see that a situation very similar to that which led to Equation 4.2 now holds. However, instead of having to manipulate Equation 4.3 into a form amenable to iterative solution, we merely note that it can be rewritten:

$$l_{ik} \leq l_{ik} * \prod_{j=1}^{n} \left[ \sum_{p=1}^{m} (l_{jp} * \Lambda_{ij}(k,p)) \right] \quad \forall i \text{ and } k \tag{4.4}$$

since the $l_{ik}$ on the right-hand side is independent of $j$ and $p$. It is clear that if Equation 4.4 does not hold, it can be made to hold by setting $l_{ik}$ equal to the value on the right-hand side. This is, in fact, equivalent to discarding label $k$ for $u_i$ if at some neighbor $u_j$ there does not exist a compatible label. If the $l_{ik}$'s, $k = 1, m$ are now gathered together in vector form:

$$\begin{bmatrix} l_{i1} \\ l_{i2} \\ \vdots \\ l_{im} \end{bmatrix} \leq \begin{bmatrix} l_{i1} \\ l_{i2} \\ \vdots \\ l_{im} \end{bmatrix} * \begin{bmatrix} \sum_{p=1}^{m} (l_{1p} * \Lambda_{i1}(1,p)) \\ \sum_{p=1}^{m} (l_{1p} * \Lambda_{i1}(2,p)) \\ \vdots \\ \sum_{p=1}^{m} (l_{1p} * \Lambda_{i1}(m,p)) \end{bmatrix} * \cdots * \begin{bmatrix} \sum_{p=1}^{m} (l_{np} * \Lambda_{in}(1,p)) \\ \sum_{p=1}^{m} (l_{np} * \Lambda_{in}(2,p)) \\ \vdots \\ \sum_{p=1}^{m} (l_{np} * \Lambda_{in}(m,p)) \end{bmatrix}$$

or

$$\begin{bmatrix} l_{i1} \\ l_{i2} \\ \vdots \\ l_{im} \end{bmatrix} \leq \begin{bmatrix} l_{i1} \\ l_{i2} \\ \vdots \\ l_{im} \end{bmatrix} * \begin{bmatrix} \mathbf{L}_1 \cdot \Lambda_{i1}(1) \\ \mathbf{L}_1 \cdot \Lambda_{i1}(2) \\ \vdots \\ \mathbf{L}_1 \cdot \Lambda_{i1}(m) \end{bmatrix} * \cdots * \begin{bmatrix} \mathbf{L}_n \cdot \Lambda_{in}(1) \\ \mathbf{L}_n \cdot \Lambda_{in}(2) \\ \vdots \\ \mathbf{L}_n \cdot \Lambda_{in}(m) \end{bmatrix}$$

or

$$\mathbf{L}_i \leq \mathbf{L}_i * \prod_{j=1}^{n} \left( \{[\mathbf{L}_j] \times [\Lambda_{ij}(1) \ldots \Lambda_{ij}(m)]\}^t \right) \tag{4.5}$$

Let

$$\mathbf{P} = [\mathbf{P}_1 \mathbf{P}_2 \mathbf{P}_3 \ldots \mathbf{P}_n]$$

where the column vector:

$$\mathbf{P}_i = \prod_{j=1}^{n} \left( \{[\mathbf{L}_j] \times [\Lambda_{ij}(1) \ldots \Lambda_{ij}(m)]\}^t \right)$$

in Equation 4.5. Gathering together the $\mathbf{L}_i$'s, $i = 1, n$, we have:

$$\mathbf{L} \leq \mathbf{L} * \mathbf{P}$$

This formulation emphasizes the relation to classical relaxation.

We will demonstrate these ideas by continuing the region coloring example. We have then, according to Equation 4.4,

For $i = 1$ and $k = 1$ :

$$\begin{aligned} l_{11} \leq\ & l_{11} * [l_{11} * \Lambda_{11}(1,1) + l_{12} * \Lambda_{11}(1,2) + l_{13} * \Lambda_{11}(1,3)] \\ & * [l_{21} * \Lambda_{12}(1,1) + l_{22} * \Lambda_{12}(1,2) + l_{23} * \Lambda_{12}(1,3)] \\ & * [l_{31} * \Lambda_{13}(1,1) + l_{32} * \Lambda_{13}(1,2) + l_{33} * \Lambda_{13}(1,3)] \end{aligned}$$

$$\begin{aligned}\Longrightarrow 1 &\leq 1*[1*1+0*0+0*0]*[0*0+0*0+1*1] \\ &\quad *[1*0+1*1+1*1]\end{aligned}$$

$$\Longrightarrow 1 \leq 1 \quad \text{which is true}$$

This says that the color red is all right for Region 1. To determine if the color red is possible for Region 2, we must find $l_{21}$.

For $i = 2$ and $k = 1$ :

$$\begin{aligned}l_{21} &\leq l_{21}*[l_{11}*\Lambda_{21}(1,1)+l_{12}*\Lambda_{21}(1,2)+l_{13}*\Lambda_{21}(1,3)] \\ &\quad *[l_{21}*\Lambda_{22}(1,1)+l_{22}*\Lambda_{22}(1,2)+l_{23}*\Lambda_{22}(1,3)] \\ &\quad *[l_{31}*\Lambda_{23}(1,1)+l_{32}*\Lambda_{23}(1,2)+l_{33}*\Lambda_{23}(1,3)]\end{aligned}$$

$$\begin{aligned}\Longrightarrow 1 &\leq 1*[1*0+0*0+0*0]*[1*1+1*0+1*0] \\ &\quad *[0*0+0*0+1*1]\end{aligned}$$

$$\Longrightarrow 1 \leq 1*0$$

$$\Longrightarrow 1 \leq 0 \quad \text{which is false}$$

Thus, $l_{21}$ must be set to zero. Likewise, for $i = 2$ and $k = 3$, $l_{23}$ is set to zero, and blue is not a possible label for Region 2. Finally,

For $i = 2$ and $k = 2$ :

$$\begin{aligned}l_{22} &\leq l_{22}*[l_{11}*\Lambda_{21}(2,1)+l_{12}*\Lambda_{21}(2,2)+l_{13}*\Lambda_{21}(2,3)] \\ &\quad *[l_{21}*\Lambda_{22}(2,1)+l_{22}*\Lambda_{22}(2,2)+l_{23}*\Lambda_{22}(2,3)] \\ &\quad *[l_{31}*\Lambda_{23}(2,1)+l_{32}*\Lambda_{23}(2,2)+l_{33}*\Lambda_{23}(2,3)]\end{aligned}$$

$$\begin{aligned}\Longrightarrow 1 &\leq 1*[1*1+0*0+0*0]*[1*0+1*1+1*0] \\ &\quad *[0*0+0*0+1*1]\end{aligned}$$

$$\Longrightarrow 1 \leq 1 \quad \text{which is true}$$

We see then that the values of $l_{11}, l_{22}$, and $l_{33}$ are not affected by the change of $l_{21}$ and $l_{23}$ to zero. In fact, the system of equations stabilizes after the change of $l_{21}$ and $l_{23}$, and the result is $l_{11} = l_{22} = l_{33} = 1$, while all other hypotheses are zero. Thus, the only consistent labeling is to label Regions 1, 2, and 3 the colors red, green, and blue, respectively.

## 4.3 Discussion

The parallel iterative discrete relaxation algorithm proposed in Reference 67 is achieved by repeating:

$$\mathbf{L} := \mathbf{L} * \mathbf{P} \tag{4.6}$$

until **L** does not change value. The form of Equation 4.6 brings out clearly:

- The parallel iterative algorithm corresponds to the classical point Jacobi method of relaxation; the Gauss-Seidel method can also be used by taking advantage of already updated unknowns, i.e., asynchronous updating is possible.

This formulation of discrete relaxation permits greater insight into the nature and use of discrete relaxation. Namely, one important insight gained by studying this formulation is that a single underlying architecture can be used to perform both classical and Boolean relaxation where the only essential difference is at the lowest level of the particular operations performed, i.e., arithmetic or logical. Given that many low-level vision operations can be performed in terms of classical relaxation, and that certain symbolic processes can be set up as a symbolic relaxation process, this could provide a possible underlying mechanism for both forms of processing.

In comparing classical relaxation to Boolean relaxation, we see that in classical relaxation local constraints along with the boundary conditions are sufficient to determine a unique global fixed point for the system. The initial vector affects the speed of convergence, but not the final answer. On the other hand, in the Boolean relaxation scheme, once a label is zero, it will stay zero. Moreover, the binary constraints do not necessarily constrain the system to a unique fixed point. In fact, the fixed point achieved by the algorithm is that consistent labeling which requires the elimination of the fewest labels from the initial set, and even if nonempty, does not in fact guarantee an unambiguous labeling. In general, the process does, however, significantly reduce the search space.

Another major difference between classical relaxation and discrete relaxation is that known boundary values are introduced in the classical updating rule:

$$\mathbf{L} \quad := \quad \mathbf{L} * \mathbf{P} + \mathbf{B}$$

where **B** is a constant-valued vector. This form actually can be used meaningfully with discrete relaxation, too. For example, if the labels of certain objects are known, then by or'ing them repeatedly (i.e., the Boolean "+"), other labels can be guaranteed support from these "absolute" labels, even though these "absolute" labels themselves lack relative support.

Since most of the classical theory developed by Varga is for real-valued solutions, it is too general for direct application when the matrixes under consideration are Boolean-valued. For a more appropriate theory for the solution of Boolean-valued simultaneous equations, see Rudeanu[69]. An approach permitting more general mapping functions has been described by Robert[66], and he has investigated the nature of general discrete iterative processes.

It should be pointed out that Hummel and Zucker[43] have shown that the more general stochastic relaxation labeling process can be nicely formulated in the framework of variational calculus. They use the notion of consistency to tie together discrete and stochastic relaxation. The definition they give counts the number of supporting neighbors rather than simply determining in a yes/no manner whether sufficient support exists as described here.

The generalization of discrete relaxation to encompass $n$-ary relations has been studied by several authors[7,17,52]. In the context of Equation 4.4, this merely involves and'ing all $n$ labels and using an $n$-dimensional matrix for the consistency relation.

# Chapter 5

# Networks of Constraints

We have seen that the consistent labeling problem is computationally hard, and indeed, we show below that it is in the class of NP-complete problems. Thus, direct solutions like generate and test, theorem proving, and backtracking suffer from the combinatorics of the relations of the labels. However, it is often the case that certain preprocessing operations (like discrete relaxation) can eliminate a good many potential labels before the search for complete solutions begins. In this chapter, we describe the labeling problem in terms of networks of constraints, where the contraints can be described as either predicates or relations. This formulation of the problem is due to Montanari[59] and has motivated much subsequent work toward finding efficient algorithms. It can be shown that a wide variety of constraint satisfaction problems can be posed in this framework: discovering relational homomorphisms between algebraic structures, solving cryptarithmetic problems, determining satisfaction of wff's in the propositional calculus and determining correct interpretations for lines in scenes of blocks[21,30,31].

## 5.1 Introduction

We now turn to Montanari's formulation of the problem[59]. In the network view of the discrete labeling problem, the variables (or objects, $u_i$) are the nodes of a directed graph. A constraint betweeen two variables can be represented in complete generality by a relation between two sets. If $\Lambda_i$ and $\Lambda_j$ are sets of labels, then the characteristic function $F$ can be used to represent relation $R_{ij}$:

$$F : \Lambda_i \times \Lambda_j \rightarrow \{0,1\} \ni F(\lambda_1, \lambda_2) = 1 \Leftrightarrow (\lambda_1, \lambda_2) \in R_{ij}$$

If $\Lambda_i$ has $m$ elements and $\Lambda_j$ has $n$ elements, then $2^{mn}$ different relations exist between $\Lambda_i$ and $\Lambda_j$. An $m \times n$ matrix whose entries are just $F(\lambda_1, \lambda_2)$ can be used to represent the relation. For example, let $\Lambda_i = \Lambda_j = \{1,2,3\}$, and let $R_{ij} = \{(1,2),(1,3),(2,1),(3,2),(3,3)\}$. Then $R_{ij}$ can be represented as a matrix:

$$\mathbf{R}_{ij} = \begin{bmatrix} 0 & 1 & 1 \\ 1 & 0 & 0 \\ 0 & 1 & 1 \end{bmatrix}$$

Negation, union, intersection, and a partial ordering relation of set inclusion can be defined in a straightforward way. Relations form a complete lattice with greatest element the matrix of all ones, denoted by **1** and least element the zero matrix, denoted by **0**. Union and intersection act as sup and inf, respectively. Once the relations are expressed as binary matrixes, it is easy to define some useful new relations:

- *Simultaneous relations*: Expressed as intersection (Boolean and). Given $\mathbf{R}'_{ij}$ and $\mathbf{R}''_{ij}$, then the resulting simultaneous relation is defined as: $\mathbf{R}_{ij}(l,m) := \mathbf{R}'_{ij}(l,m) * \mathbf{R}''_{ij}(l,m)$.

- *Composition of relations*: Given $\mathbf{R}_{ij}$ and $\mathbf{R}_{jk}$ their composition is denoted as:

$$\mathbf{R}_{ik} := \mathbf{R}_{ij} \circ \mathbf{R}_{jk}$$

  which is defined as:

$$\mathbf{R}_{ik}(r,s) \quad := \quad \sum_{t=1}^{m} \mathbf{R}_{ij}(r,t) * \mathbf{R}_{jk}(t,s)$$

  Furthermore, composition takes precedence over intersection.

- *Induced relations*: These are defined for $\mathbf{R}_{ij}$ as:

$$\mathbf{R}'_{ij} \quad := \quad \mathbf{R}_{ij} * \mathbf{R}_{ik} \circ \mathbf{R}_{kj} \quad \text{for some } k$$

Note that when a relation is expressed as a set, we use $R$, and when in matrix form, we use $\mathbf{R}$. We would like to consider constraints between more than two variables at a time, but the amount of information grows exponentially with the number of variables. This poses practical difficulties since an $n$-ary relation $R$ is any subset of $X = \Lambda_1 \times \Lambda_2 \times \cdots \times \Lambda_n$, and if each $\Lambda_i$ has $m$ elements, then there are a total of $2^{m^n}$ $n$-ary relations. However, projections of $n$-ary relations to networks of binary relations present a useful alternative. A *network of binary relations* $N$, comprises two things:

1. A set of sets $X = \{\Lambda_1, \ldots, \Lambda_n\}$
2. A relation $R_{ij}$ from every set $\Lambda_i$ to every set $\Lambda_j$, for $i, j = 1, \ldots, n$.

The nodes of the network are the elements of $X$, and the relations $R_{ij}$ can be viewed as labeling the edges between nodes. Given $X$ as above, there are $2^{(m^2)[n(n+1)/2]}$ networks of binary relations, assuming $R_{ij} = R_{ji}$. Thus, it is clear that there are many more distinct $n$-ary relations than binary relation networks.

A given network, $N$, represents the $n$-ary relation:

$$\begin{aligned} R \quad = \quad & \{\lambda \mid \lambda = (\lambda_1, \ldots, \lambda_n) \in X = \Lambda_1 \times \Lambda_2 \times \cdots \times \Lambda_n \\ & \text{and } \forall i, j, \mathbf{R}_{ij}(\lambda_i, \lambda_j) = 1\} \end{aligned} \tag{5.1}$$

That is, the $n$-tuple $\lambda$ is in $R$ if all its 2-D projections satisfy the binary constraints. The space required to store a general $n$-ary relation is on the order of $m^n$, while each binary relation network takes on the order of $n^2m^2$.

Given a general $n$-ary relation, $R$, we can also define a binary relation network to represent it by taking the binary projection of each pair of variables:

$$\mathbf{R}_{ij}(\lambda_i, \lambda_j) \quad = \quad 1 \text{ if } \exists \lambda = (\lambda_1, \ldots, \lambda_i, \ldots, \lambda_j, \ldots, \lambda_n) \in R \tag{5.2}$$

When the $n$-ary relation $R$ is given in column form as a matrix, the $R_{ij}$'s are found by defining the 2 by $n$ matrix $\mathbf{M}$ as:

$$\begin{aligned}\mathbf{M}(1,k) &:= R(i,k)\\ \mathbf{M}(2,k) &:= R(j,k) \quad k=1,n\end{aligned}$$

Note that $\mathbf{M}$ only takes on values in the range 1 to $m$, the number of labels. We have then that:

$$\mathbf{R}_{ij}(\mathbf{M}(1,k),\mathbf{M}(2,k)) = 1 \quad k=1,n$$

Indeed, it can be shown that the network obtained by this method is a sort of least upper bound approximation to $R$. A point on notation: given network $N$ generated from relation $R$, we represent its binary relations by $R_{ij}$; e.g., for network $N'$ from relation $R'$ we use $R'_{ij}$, etc.

**Theorem 5.1** *(i) The $n$-ary relation $R'$ represented by the network defined by Equation 5.2 contains $R$, the original relation (that is, $R \subseteq R'$). (ii) Moreover, there does not exist a network $N''$ which induces an $n$-ary relation $R'' \ni R \subseteq R'' \subseteq R'$.*

Proof:
(i) $R \subseteq R'$. Let $\lambda = (\lambda_1, \lambda_2, \ldots, \lambda_n) \in R$. Then by Equation 5.2, $R_{12}(\lambda_1,\lambda_2) = 1, R_{13}(\lambda_1,\lambda_3) = 1$, etc., and in general $R_{ij}(\lambda_i,\lambda_j) = 1$. Then by Equation 5.1, $(\lambda_1, \lambda_2, \ldots, \lambda_n) \in R'$.
(ii) Show that it is not the case that $R \subseteq R'' \subset R'$. Suppose there exists $N''$ and $R'' \ni R \subseteq R'' \subset R'$. Then there exists $\lambda = (\lambda_1, \ldots, \lambda_n) \ni \lambda \in R'$ but $\lambda \notin R''$. Then there exists a pair $(\lambda_i, \lambda_j)$ of $\lambda \ni R''_{ij}(\lambda_i,\lambda_j) = 0$, while $R'_{ij}(\lambda_i, \lambda j) = 1$. But $R'_{ij}(\lambda_i,\lambda_j) = 1$ implies there exists $\lambda' \in R \ni (\lambda_i,\lambda_j)$ is in $\lambda'$. Then $\lambda' \notin R''$ and $R \not\subseteq R''$. But this is a contradiction. **Q.E.D.**

Let us now look at an example of a relation $R$, its corresponding network, $N$, and the relation $R'$ represented by $N$. Let:

$$\begin{aligned}U &= \{u_1,u_2,u_3\}\\ \Lambda_1 &= \Lambda_2 = \{1,2,3\}\\ \Lambda_3 &= \{1,2\}\\ R &= \{(1,1,1),(2,2,1),(2,1,2)\}\end{aligned}$$

or in column format:

$$\mathbf{R} = \begin{bmatrix}1&2&2\\1&2&1\\1&1&2\end{bmatrix}$$

The binary relation network is defined by:

$$\text{Since } \mathbf{M}_{12} = \begin{bmatrix}1&2&2\\1&2&1\end{bmatrix} \text{ then } \mathbf{R}_{12} = \begin{bmatrix}1&0\\1&1\end{bmatrix}.$$

$$\text{Since } \mathbf{M}_{23} = \begin{bmatrix}1&2&1\\1&1&2\end{bmatrix} \text{ then } \mathbf{R}_{23} = \begin{bmatrix}1&1\\1&0\end{bmatrix}.$$

$$\text{Since } \mathbf{M}_{13} = \begin{bmatrix}1&2&2\\1&1&2\end{bmatrix} \text{ then } \mathbf{R}_{13} = \begin{bmatrix}1&0\\1&1\end{bmatrix}, \text{ and } \mathbf{R}_{11} = \mathbf{R}_{22} = \mathbf{R}_{33} = \mathbf{I}.$$

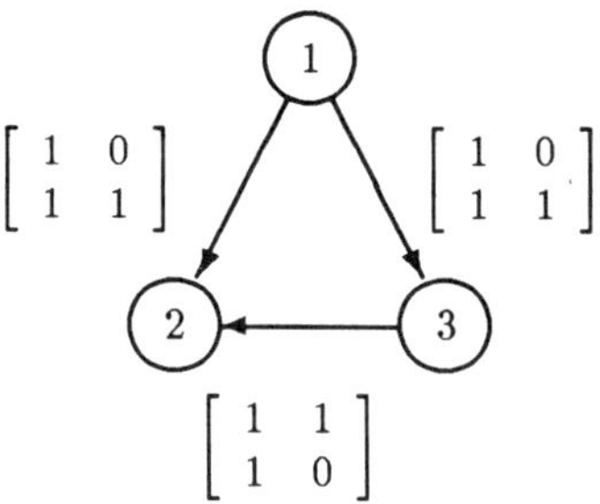


Figure 5.1: The binary relation network.

The network is shown in Figure 5.1. The induced $n$-ary relation from this network is:

$$\mathbf{R}' = \begin{bmatrix} 1 & 2 & 2 & 2 \\ 1 & 1 & 1 & 2 \\ 1 & 1 & 2 & 1 \end{bmatrix}$$

Note that $\mathbf{R}$ is not representable as a binary relation network since the induced relation $\mathbf{R}'$ does not equal $\mathbf{R}$. For a relation which is representable by a binary relation network, consider the following:

$$\mathbf{R} = \begin{bmatrix} 1 & 2 & 3 \\ 1 & 2 & 3 \\ 1 & 2 & 3 \end{bmatrix}$$

Then the binary relations are all derived from:

$$\mathbf{M} = \begin{bmatrix} 1 & 2 & 3 \\ 1 & 2 & 3 \end{bmatrix}$$

and

$$\mathbf{R}_{12} = \mathbf{R}_{23} = \mathbf{R}_{13} = \begin{bmatrix} 1 & 0 & 0 \\ 0 & 1 & 0 \\ 0 & 0 & 1 \end{bmatrix}$$

From which it is clear the induced relation is:

$$R' = \{(1,1,1),(2,2,2),(3,3,3)\}$$

We define an ordering relation between networks by:

$$N' \subseteq N'' \quad \Leftrightarrow \mathbf{R}'_{ij} \leq \mathbf{R}''_{ij} \;\; i,j = 1,n$$

The set of all networks with $n$ vertexes is a lattice under $\subseteq$. The least element occurs when $\mathbf{R}_{ij} = \mathbf{0}$ and the greatest element occurs when $\mathbf{R}_{ij} = \mathbf{1}(i \neq j)$ and $\mathbf{R}_{ii} = \mathbf{I}$. Union and intersection are then defined in a straightforward way in terms of the $\mathbf{R}_{ij}$'s.

An $n$-ary relation may be represented by several networks. Consider $N_1$ and $N_2$ given in Figure 5.2. Two networks which represent the same $n$-ary relation are called *equivalent*. If $N'$ and $N''$ are equivalent, we denote this as $N' \equiv N''$. In fact, a minimal network exists for any set of equivalent networks.

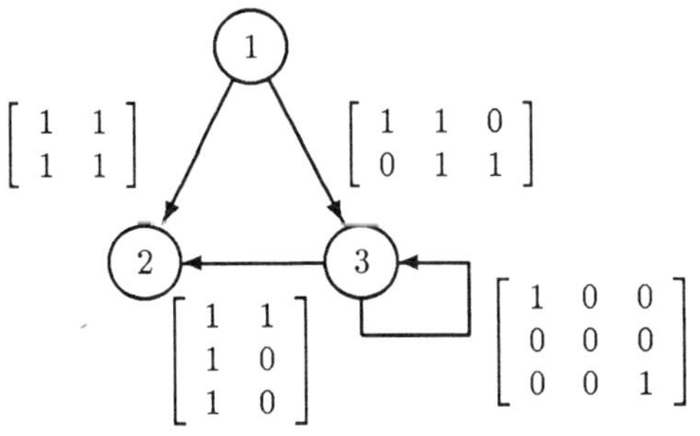

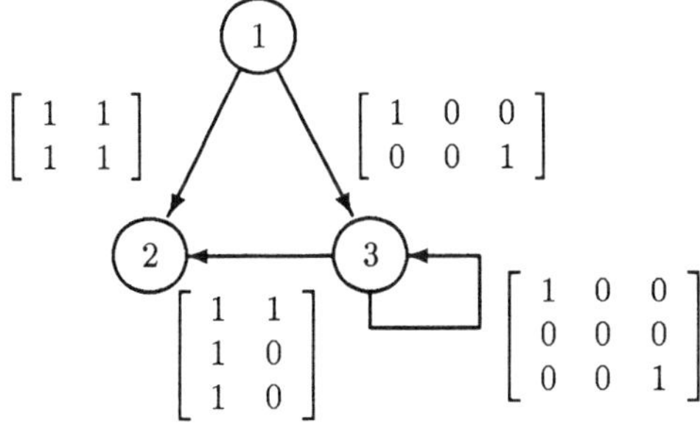

Figure 5.2: Equivalent networks.

**Theorem 5.2** *Given a set of networks which represent the same relation $R$, we let $N_R$ be the equivalence class defined by that set. If $N' \in N_R$ and $N'' \in N_R$, then $N = N' * N'' \in N_R$. The application of the projection formula of equation 5.2 to a relation $R$ results in the minimal network representing $R$.*

**Proof**:
Suppose that $N \subseteq N'$ and $N \subseteq N''$. Then, an $n$-tuple $\lambda$ satisfying $N$ will also satisfy $N'$ and $N''$. This is true since $N \subseteq N'' \Rightarrow R \subseteq R'$. On the other hand, $\lambda$ satisfies $N$ if it satisfies $N'$ and $N''$. If $R'_{ij}(\lambda_1, \lambda_2) = 1$ and $R''_{ij}(\lambda_1, \lambda_2) = 1$, then since $N = N' * N''$ (by the definition of "and"), we have that $R_{ij}(\lambda_1, \lambda_2) = 1$. To see that the network obtained from equation 5.2 is minimal, note that for any pair of labels $(\lambda_1, \lambda_2)$, if $R_{ij}(\lambda_1, \lambda_2)$ is set to zero, the underlying relation $R$ is changed. **Q.E.D.**

**Theorem 5.3** *A necessary and sufficient condition for a network $N$ to be minimal is that if $\exists i, j, \lambda'$ and $\lambda'' \ni R_{ij}(\lambda', \lambda'') = 1$, then an $n$-tuple $\lambda$ satisfying $N$ exists $\ni \lambda = (\lambda_1, \ldots, \lambda_i, \ldots, \lambda_j, \ldots, \lambda_n) \in R$.*

**Proof**:
($\Rightarrow$) If $N$ is minimal and $R_{ij}(\lambda', \lambda'') = 1$, then an $n$-tuple $\lambda = (\lambda_1, \ldots, \lambda_i, \ldots, \lambda_j, \ldots$ $\lambda_n)$ with $\lambda \in R$ must exist or else $N'$ with $R'_{ij}$ having $R'_{ij}(\lambda', \lambda'') = 0$ is equivalent and smaller.

($\Leftarrow$) From the assumption, no pair $(\lambda, \lambda')$ can be deleted from any binary relation $R_{ij}$ and keep an equivalent network. This means that, $N$ is minimal. **Q.E.D.**

A minimal network of constraints is perfectly explicit: $M_{ij}$, the direct constraint between variables $u_i$ and $u_j$, tells everything there is to know. Unfortunately, generating the $n$-ary relation $R$ from the original $N$ and then $M$ from $R$ with Equation 5.2 is almost always infeasible. The *central problem* in using networks of constraints to represent an $n$-ary relation is to find the minimal network equivalent to the given network. No algorithm is known other than complete enumeration. Results have been obtained for restricted classes of networks. The main practical result is an algorithm for finding all the implicit binary constraints between pairs of objects, i.e., the closure of a relation. Then the closure of $R$ is the largest closed network not larger than $R$ but equivalent to $R$; that is, $R$ and the closure of $R$ represent the same $n$-ary relation. We will formulate this problem more precisely in the next section.

## 5.2 Network Consistency

Another approach to using constraints has its origins in scene analysis. Given a segmentation of a scene, the objects produced by the segmentation may not have a clear interpretation. That is, some type of analysis is performed on each object, but without knowledge of other objects in a scene, disambiguation of the object is not possible. More precisely, given a prototypical scene, relations can be defined from the known constraints between objects in the scene. These relations can be used to disambiguate objects in an unknown scene by ruling out interpretations which do not satisfy all (or some) of the required constraints with the interpretations of other objects in the scene. For example, the objects in an airplane (e.g., wings, nose, and fuselage) are constrained in the sense that the nose must be adjacent to two wings and a fuselage, whereas the fuselage is also adjacent to a tail. Finding only the interpretations satisfying the constraints is the *scene labeling problem*.

Recall that the labeling problem is defined as follows: Let:

1. $U = \{u_1, \ldots, u_n\}$ be a set of *objects*.
2. $\Lambda = \{\lambda_1, \ldots, \lambda_m\}$ be a set of *labels*.
3. $L : A^k \rightarrow \Lambda^k$ be a *labeling* $(k = 1, n)$.
4. $P \subseteq (U \times \Lambda)^l$ be an *order-l compatibility* relation.
5. If $L(u_1, \ldots, u_k) = (\lambda_1, \ldots, \lambda_k)$, then $L$ is *consistent* iff $\forall l$-tuple $(u_{i_1}, \ldots, u_{i_l})$ of $u_i$'s in $(u_1, \ldots, u_k)$, $(u_{i_1}, \lambda_{i_1}, \ldots, u_{i_l}, \lambda_{i_l}) \in P$.

A *globally consistent labeling* is a labeling for which $k = n$. The *labeling problem* is to find a globally consistent labeling for a given set of objects, labels, and relations. First, we prove that the consistent labeling problem is computationally difficult. Our proof follows that of Haralick et al.[30].

**Theorem 5.4** *The labeling problem is NP-complete.*

**Proof**:
First, we show that the problem can be solved in polynomial time by a nondeterministic Turing machine. To do this, simply choose a labeling nondeterministically

and verify its global consistency. Given an order-$l$ compatibilility relation, it is necessary to check that every $l$-tuple of objects is indeed labeled as allowed by the constraints. Since $l$ is fixed, the number of labelings is just $\binom{n}{l}$ which is bounded by $n^l$.

Next, we show that if the labeling problem can be solved, then so can the colorability problem. If in a graph $G = (V, E)$ it is possible, given $k$ colors, to color the vertexes of the graph such that no two adjacent vertexes have the same color, then the graph is $k$-colorable. To map this onto the labeling problem, let $\Lambda = \{1, 2, \ldots, k\}$ represent the $k$ colors, $U = V$, and let

$$P \quad = \quad \{(u_1, \lambda_1, u_2, \lambda_2) \in (U \times \Lambda)^2 \mid (u_1, u_2) \in E \text{ implies } \lambda_1 \neq \lambda_2\}$$

Suppose that $(\lambda_1, \ldots, \lambda_n)$ is a globally consistent labeling; then every $(u_i, \lambda_i, u_j, \lambda_j) \in P$ implies that $(u_i, u_j) \in E$, and therefore, that $\lambda_i \neq \lambda_j$. This means that no two adjacent vertexes have the same color. **Q.E.D.**

As we have seen, rather than solve the consistent labeling problem directly on $n$-ary relations, we will focus on producing the transitive closure on the binary relations of the corresponding network. We need some more definitions.

**Definition 5.1** *A labeling is:*

1. Node consistent *if it is order-1 compatible*

2. Arc consistent *if it is order-2 compatible*

3. Path consistent *if it is order-2 compatible and if* $\forall$ *path through* $m+1$ *nodes* $(i_0, \ldots, i_m) \forall \lambda \in \Lambda_{i_0}$ *and* $\lambda' \in \Lambda_{i_m} \ni (\lambda, \lambda') \in R_{i_0 i_m} \exists z_j \in \Lambda_j \ni (\lambda, z_j) \in R_{i_{j-1} i_j} (j = 1, m)$

Recall that unary relations are described by loops on the nodes and that binary relations are directed arcs between the nodes. For example, consider the graph shown in Figure 5.3. Then, $\Lambda_1 = \Lambda_2 = \{1, 2, 3, 4\}$ and $\Lambda_3 = \Lambda_4 = \Lambda_5 = \{1, 2\}$; the relation is "$<$" (less than). A solution is then a labeling $(\lambda_1, \lambda_2, \lambda_3, \lambda_4, \lambda_5) \ni \lambda_i \in \Lambda_i$ which satisfies the "$<$" arcs. Generate and test examines 128 possibilities. A resolution theorem prover requires a combinatorial search for resolvents (see Figure 5.4). Remember that "+" stands for the Boolean "or" operation, "*" stands for the Boolean "and" operation, and "$\neg$" stands for negation, Line 16 gives the negation of the assertion:

$$\exists x_1 x_2 x_3 x_4 x_5 ((x_1 < x_3) * (x_2 < x_3) * (x_4 < x_3) * (x_5 < x_3) * (x_5 < x_4))$$

which defines the constraint structure. Essentially, a resolution theorem prover makes all possible substitutions which unify the facts in Lines 1 to 15 with the negated theorem in Line 16. Alternatively, backtrack search can be used. A trace of a backtrack search is given in Figure 5.5.

Let us now consider node, arc, and path consistency in more detail. Suppose there is a unary predicate on $a_1$:

$$Value(a_1) < 4$$

i.e., the value of $a_1$ must be less than 4. Then, $N$ is not node consistent since $4 \in \Lambda_1$. Moreover, $N$ is not arc consistent since for the value 2 at node $a_3$, there is

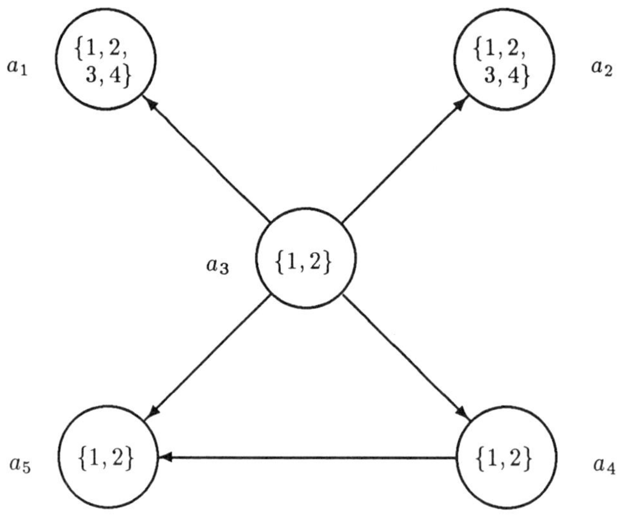


Figure 5.3: Example constraint network.

1. $\neg(1 < 1)$
2. $\neg(2 < 2)$
3. $\neg(3 < 3)$
4. $\neg(4 < 4)$
5. $1 < 2$
6. $1 < 3$
7. $1 < 4$
8. $2 < 3$
9. $2 < 4$
10. $3 < 4$
11. $(x_1 = 1) + (x_1 = 2) + (x_1 = 3) + (x_1 = 4)$
12. $(x_2 = 1) + (x_2 = 2) + (x_2 = 3) + (x_2 = 4)$
13. $(x_3 = 1) + (x_3 = 2)$
14. $(x_4 = 1) + (x_4 = 2)$
15. $(x_5 = 1) + (x_5 = 2)$
16. $\neg(x_1 < x_3) + \neg(x_2 < x_3) + \neg(x_4 < x_3) + \neg(x_5 < x_3) + \neg(x_5 < x_4)$

Figure 5.4: Resolution theorem proving axioms.

| $a_1$ | $a_1a_2$ | $a_1a_2a_3$ | $a_1a_2a_3a_4$ | $a_1a_2a_3a_4a_5$ |
|---|---|---|---|---|
| 1 | 11 | 111 | | |
| | | 112 | | |
| | 12 | 121 | | |
| | | 122 | | |
| | 13 | 131 | | |
| | | 132 | | |
| | 14 | 141 | | |
| | | 142 | | |
| 2 | 21 | 211 | | |
| | | 212 | | |
| | 22 | 221 | 2211 | |
| | | | 2212 | 22121 |
| | | | | 22121 |
| | | 222 | | |
| | 23 | 231 | 2311 | |
| | | | 2312 | 23121 |
| | | | | 23122 |
| | 24 | 241 | 2411 | |
| | | | 2412 | 24121 |
| | | | | 24122 |
| | | 242 | | |
| 3 | ...... | | | |

Figure 5.5: Backtrack trace for problem with constraint network.

no consistent value at $a_5$. Finally, $N$ is not path consistent; consider the path from node $a_3$ to node $a_5$ through node $a_4$. If we take label 1 at $a_3$ and label 2 at $a_5$, they are consistent, but there is no label at $a_4$ which is consistent simultaneously with 1 at $a_3$ and 2 at $a_5$.

A more complicated problem is the $N$-queens problem. The goal is to place $n$ queens on a chessboard such that no two queens are in the same row, column, or diagonal (see Figure 5.6). To be more concrete, let $n = 4$. For this problem, we can define the constraint relations in terms of the binary matrixes $\mathbf{R}_{ij}$ defined as:

$$\mathbf{R}_{ij}(l,m) = \begin{cases} 0 & \text{if queen } i \text{ in row } l \text{ prohibits queen } j \text{ in row } m \\ 1 & \text{otherwise} \end{cases}$$

The queens are then assigned one to a column, and their possible labels are the row positions in the column. We have then:

$$\mathbf{R}_{11} = \mathbf{R}_{22} = \mathbf{R}_{33} = \mathbf{R}_{44} = \begin{bmatrix} 1 & 0 & 0 & 0 \\ 0 & 1 & 0 & 0 \\ 0 & 0 & 1 & 0 \\ 0 & 0 & 0 & 1 \end{bmatrix} \text{ and } \mathbf{R}_{ij} = \mathbf{R}_{ji}^t$$

The constraint matrixes for the 4-queens problem ($i \neq j$) are:

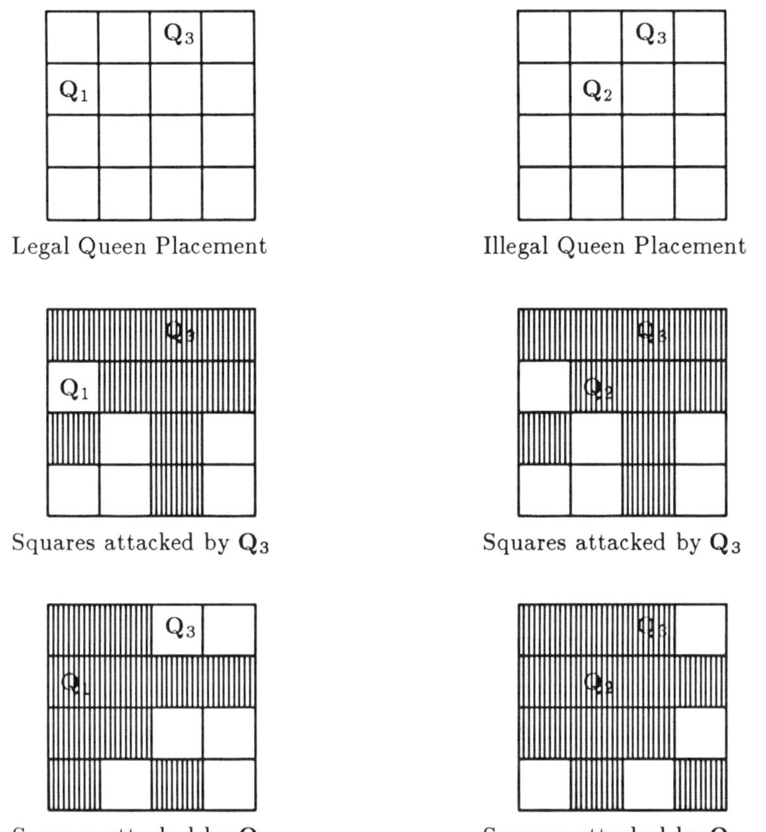


(a) Queens satisfying constraints. (b) Queens not satisfying constraints.

Figure 5.6: Legal and illegal queen arrangements.

$$\mathbf{R}_{12} = \mathbf{R}_{23} = \mathbf{R}_{34} = \begin{bmatrix} 0 & 0 & 1 & 1 \\ 0 & 0 & 0 & 1 \\ 1 & 0 & 0 & 0 \\ 1 & 1 & 0 & 0 \end{bmatrix}$$

$$\mathbf{R}_{13} = \mathbf{R}_{24} = \begin{bmatrix} 0 & 1 & 0 & 1 \\ 1 & 0 & 1 & 0 \\ 0 & 1 & 0 & 1 \\ 1 & 0 & 1 & 0 \end{bmatrix}$$

$$\mathbf{R}_{14} = \begin{bmatrix} 0 & 1 & 1 & 0 \\ 1 & 0 & 1 & 1 \\ 1 & 1 & 0 & 1 \\ 0 & 1 & 1 & 0 \end{bmatrix}$$

while Figure 5.7 shows the binary relation network for this problem. Figure 5.8 describes the role that induced relations play in discovering consistency. Recall

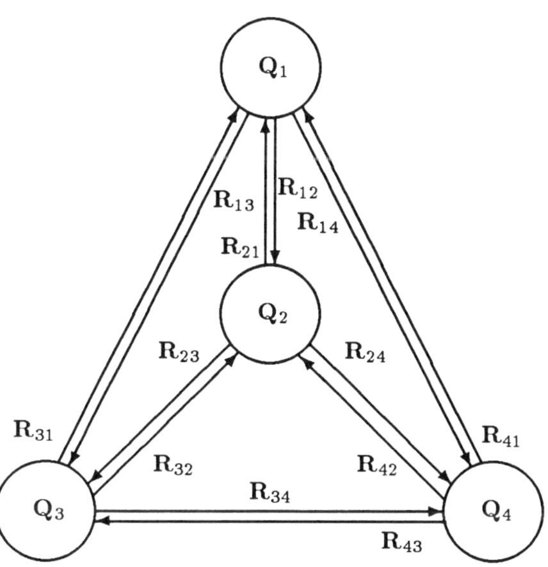


Figure 5.7: Graph corresponding to the 4-queens problem.

that, in general, an induced relation is given by $\mathbf{R}'_{ij} := \mathbf{R}_{ij} * \mathbf{R}_{ik} \circ \mathbf{R}_{kj}$. We can use induced relations to eliminate impossible queen placements which are not already explicit in the binary relation. In this case, it is clear that the placement of $Q_1$ in row 1 and $Q_3$ in row 4 eliminates all possible placements of $Q_2$. Even though $\mathbf{R}_{13}(1,4) = 1$, we find that $\mathbf{R}_{12} \circ \mathbf{R}_{23}(1,4) = 0$. So we can induce that $\mathbf{R}'_{13}(1,4) = 0$, where $\mathbf{R}'_{13}$ is the induced relation and $k = 2$. That is,

$$\mathbf{P}_{12} = \begin{bmatrix} 0 & 0 & 1 & 1 \\ 0 & 0 & 0 & 1 \\ 1 & 0 & 0 & 0 \\ 1 & 1 & 0 & 0 \end{bmatrix}$$

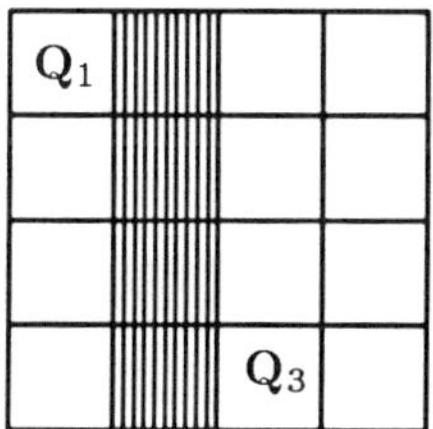


Figure 5.8: An induced relation on the 4-queens problem.

$$\mathbf{P}_{23} = \begin{bmatrix} 0 & 0 & 1 & 1 \\ 0 & 0 & 0 & 1 \\ 1 & 0 & 0 & 0 \\ 1 & 1 & 0 & 0 \end{bmatrix}$$

$$\mathbf{P}_{12} \circ P_{23} = \begin{bmatrix} 1 & 1 & 0 & 0 \\ 1 & 1 & 0 & 0 \\ 0 & 0 & 1 & 1 \\ 0 & 0 & 1 & 1 \end{bmatrix}$$

$$\mathbf{P}_{13} = \begin{bmatrix} 0 & 1 & 0 & 1 \\ 1 & 0 & 1 & 0 \\ 0 & 1 & 0 & 1 \\ 1 & 0 & 1 & 0 \end{bmatrix}$$

$$\mathbf{P}'_{13} = \begin{bmatrix} 0 & 1 & 0 & 0 \\ 1 & 0 & 0 & 0 \\ 0 & 0 & 0 & 1 \\ 0 & 0 & 1 & 0 \end{bmatrix}$$

Thus, the induced relation makes explicit certain constraints which were only implicit before. Obviously, it would be useful if all such induced relations were made explicit. We next consider how this can be achieved.

The node consistency condition consists only in checking the unary relations on the different nodes and keeping in the domain of each node values satisfying this unary constraint. Arc consistency checks the consistency of labels for each couple of nodes linked by a binary constraint and removes the labels that cannot satisfy this local condition.

Path consistency algorithms ensure that any pair of labelings $(i, \lambda)$ and $(j, \lambda')$ allowed by a direct relation is also allowed by all paths from $i$ to $j$. It has been proven that for complete graphs, path consistency is equivalent to consistency of every path of length 2; therefore, this is equivalent to checking the consistency of every triple. Each graph can always be replaced by an equivalent complete graph by adding the *true* constraint between the nodes which are not connected.

We shall examine algorithms which ensure node, arc, and path consistency. Although discovering such algorithms is not difficult, finding efficient algorithms is more involved. There are certain tradeoffs in terms of programming ease, and of course space and time complexity.

## 5.3 Node Consistency

**procedure** NC-1;
**begin**
 **for** $u_i \in U$ **do**
  **begin**
  $L_i := \Lambda_i$;
  **for** $\lambda \in L_i$ **do**
   **if** $\neg R_i(\lambda)$ **then** $L_i := L_i - \{\lambda\}$;
**end**;

**Algorithm NC-1**. Node consistency algorithm.

Node consistency is very easy to achieve. Algorithm NC-1 gives one method for achieving node consistency. (In naming the algorithms, we have followed the original usage; see References 52, 59, and 67. The complexity of this algorithm is proportional to $| U | \cdot | A |= nm$. Therefore, Algorithm NC-1 is of time complexity $O(nm)$. Moreover, it may require the evaluation of each unary relation, $R_i$.

Typically, the unary relations (or predicates) are defined in terms of some absolute values; e.g., $u_3 < 12$. But it is also useful to be able to define unary predicates in terms of the first-order statistics of measurements on the set of unknowns. For example, if the unknowns are line segments on the boundary of a shape, it may be possible to eliminate many labels for the longest or shortest line segments. Other features might include: orientation, texture, or color. As will be discussed in a later chapter, multiple, sparse, semantic relations may make the analysis task much easier than the use of a single complete relation.

In any case, the first step in achieving consistency is to ensure that node consistency holds. This oftentimes reduces quite effectively the combinatorics of the later processing. In each of the arc and path consistency algorithms presented below, we assume that the graphs are already node consistent.

## 5.4 Arc Consistency

```
procedure AC-1;
begin
  repeat
  inconsistent := false;
  for (i, j) ∈ E do
    inconsistent := (¬ support(i, j) or inconsistent);
  until ¬ inconsistent;
end;
Boolean function support(i, j);
begin
  consistent := true;
  for λ ∈ L_i do
    begin
    support := false;
    for λ' ∈ L_j do
      support := (support or R_ij(λ, λ'));
    if ¬ support then
      begin
      consistent := false;
      L_i := L_i − {λ};
      end;
    end;
  support := consistent;
end;
```

**Algorithm AC-1.** Simple arc consistency algorithm.

The essence of arc consistency is that given a pair of nodes $u_i$ and $u_j$ with associated label sets $L_i$ and $L_j$ every label $\lambda \in L_i$ is consistent with some label $\lambda' \in L_j$. Many algorithms have been proposed, and we will give most of them here. One straightforward way to encode this is as follows. Let $E = \{(i,j) \mid$ there is an arc from$u_i$ to $u_j\}$. One of the first algorithms proposed, Algorithm AC-1, checks that each arc is consistent. If an arc is found which is not consistent, that arc is made consistent and the whole process repeated. A major source of inefficiency arises when an arc is made consistent (by deleting a label at some node). All arcs are rechecked for consistency, when in fact, usually only a few other arcs could be made inconsistent by the removed label.

Observe that $\mid E \mid$ is the number of directed edges in the network. The *repeat* loop of AC-1 is executed until no inconsistent pairs of labels are found. The greatest number of iterations occurs when only one label is deleted (in the function *support*) each time through the *repeat* loop. Since there are $n$ nodes and $m$ labels per node, there are at most $nm$ iterations. Let $e = \mid E \mid$. Then, for every iteration, the *for* loop is executed $e$ times; i.e., there are $e$ calls to *support*. Each execution of *support* requires up to $m^2$ accesses to the $\mathbf{R}_{ij}$ matrix. Thus, the worst case complexity of AC-1 is $O(nem^3)$. Since $e$ is on the order of $n^2$ for general graphs, the worst case time complexity is $O(n^3m^3)$.

It is possible to organize the consistency checking in terms of a single pass through the nodes. This approach is due to Waltz[88].

```
procedure AC-2;
begin
 for i = 1, n do
   begin
   Q := {(i, j) | (i, j) ∈ E and j < i};
   Q' := {(j, i) | (j, i) ∈ E and j < i};
   while Q ≠ ∅ do
     begin
     while Q ≠ ∅ do
       begin
       remove (k, m) from Q;
       if ¬ support(k, m) then Q :=Q' ⋃{(p, k) | (p, k) ∈ E and p ≤ i and p ≠ m};
       end;
     Q := Q';
     Q' := ∅;
     end;
   end;
end;
```

**Algorithm AC-2**. A one-pass consistency algorithm.

Algorithm AC-2 makes sure that after node $i$ is introduced, all of the arcs $(k,m)$ are consistent, where $k, m \leq i$ and $k \neq m$. All the arcs leading to and from the newly introduced node must be checked. Moreover, if a label is deleted at node $k$ on arc $(k,m)$, only the arcs directed toward $k$ have to be checked.

A more general way to organize the process of AC-2 (due to Mackworth) is to maintain a queue of arcs which must be checked. If some label is deleted at node $k$ of arc $(k,m)$, then all arcs $(i,k)$ which are directed at $k$ must be added to the queue. Of course, if an arc is already on the queue, it need not be added again. Algorithm AC-3 shows this procedure. The key idea of Algorithm AC-3 is that when a label is removed from node $i$, consider only the edges $(i,j)$ because they are the only ones whose arc consistency may be affected by the change. Therefore, Algorithm AC-3 has complexity $O(em^3)$ instead of $O(enm^3)$ for the brute force Algorithm AC-1. This is shown as follows. An arc is added to $Q$ if *support*$(k, m)$ fails to be true. In this case, the number of arcs added to $Q$ is bounded by $(d_k - 1)$, where $d_k$ is the degree of node $k$. This may happen up to $m$ times per node, so the total possible number of additions to $Q$ is:

$$\sum_{k=1}^{n} m(d_k - 1) = m(e - n)$$

Since one arc is deleted from $Q$ at each iteration, the number of iterations is bounded by the initial length of $Q$ and the total number of added arcs:

$$e + m(e - n)$$

The total number of accesses to the relation matrixes is then $m^2[e + m(e - n)]$, since there can be up to $m^2$ checks for each call to *support*. Therefore, the time complexity of AC-3 is $O(m^3e)$, which for general graphs is $O(m^3n^2)$.

```
procedure AC-3;
begin
  Q := {(i,j) | (i,j) ∈ E and i ≠ j};
  while Q ≠ ∅ do
    begin
    remove (k,m) from Q;
    if ¬ support(k,m) then Q := Q ⋃{(i,k) | (i,k) ∈ E and i ≠ k and i ≠ m};
    end;
end;
```

**Algorithm AC-3**. A queue-based consistency algorithm.

Finally, it is possible to perform perfect bookkeeping of the support between labels so that no unnecessary checking occurs during consistency enforcement. This leads to AC-4 (due to Mohr and Henderson[58]), an optimal algorithm for arc consistency. The bookkeeping aspect of AC-4 results in higher storage requirements as well as a less obvious algorithm, and should be carefully analyzed when choosing among the different AC algorithms.

This improvement is based on the technical aspect of the ALICE system[51] (a general constraint satisfaction system). ALICE was designed to solve most combinatorial problems using a unified and general strategy. However, it is not possible to express in this system that we want only to eliminate labels that are locally inconsistent. Carefully looking at how ALICE runs on this problem shows that it automatically applies Algorithm AC-4. Then it starts to find a solution to the complete problem by using — loosely speaking — backtracking. In fact, it applies AC-4 at each stage of backtracking.

If we consider arc consistency intuitively, we find that it is based on the notion of support. Suppose we are considering label $\lambda$ at node $i$. As long as $\lambda$ has a minimum of support from the labels at each of the other nodes $j (j \neq i)$, $\lambda$ is considered a viable label for node $i$. But once there exists a node at which no remaining label satisfies the required relation with $\lambda$, then $\lambda$ can be eliminated as a possible label for node $i$.

Algorithm AC-4 makes this support evident by assigning a counter to each arc-label pair. Such pairs are denoted by **Counter**$[(i,j), \lambda]$ and indicate the arc from

$i$ to $j$ with label $\lambda$ at node $i$. In addition, for each label $\lambda'$ at node $j$, the set $S_{j\lambda'}$ is constructed, where $S_{j\lambda'} = \{(i,\lambda) \mid \lambda' \text{ at node } j \text{ supports } \lambda \text{ at node } i\}$; that is, if $\lambda'$ is eliminated at node $j$, then counters at $[(i,j),\lambda]$ must be decremented for each $\lambda$ supported by $\lambda'$. Finally, we use a table, **M**, to keep track of which labels have been deleted from which objects, and a list, *List*, to control the propagation of constraints. The algorithm for arc consistency is given by Algorithm AC-4. Assume node consistency has already been assured.

Step 1

```
1 M := 0; S_iλ := ∅; List := ∅;
2 for (i, j) ∈ E do
3   for λ ∈ A_i do
4     begin
5     total := 0;
6     for λ' ∈ A_j do
7       if R(i, λ, j, λ') then begin
8           total := total+1;
9           Append(S_jλ', (i, λ));
10            end
11      if total = 0 then M[i, λ] := 1;
12          else Counter[(i, j), λ] := total;
13      end;

14 initialize List from M;
```

Step 2

```
15 while List ≠ ∅ do
16   begin
17   choose (j, λ') from List and remove (j, λ') from List;
18   for (i, λ) ∈ S_jλ do
19     begin
20     Counter[(i, j), λ] := Counter[(i, j), λ] − 1;
21     if Counter[(i, j), λ] = 0 and M[i, λ] = 0
22       then begin
23         Append(List, (i, λ));
24         M[i, λ] := 1;
25         end;
26   end
27 end.
```

**Algorithm AC-4**. Optimal complexity arc consistency algorithm.

It is easy to see that Line 7 of the innermost loop of the data structure initialization part of the algorithm can be executed at most $em^2$ times. Thus, the number of elements in the sets $S_{j\lambda'}$ is on the order of $em^2$. Since Line 12 is executed at most $em$ times, the total number of counters is of the order $em$; furthermore, since the value of total is bounded by $m$, then the maximum value for a counter is $m$. Line 14 simply puts the unique $(i,\lambda)$ pairs into list form; this requires order $nm$ time. Our measure of time complexity for the algorithm is the decrementing of a counter; note that the counters and decrement lists encode in a fixed way the binary relations. Thus, this measure is consistent with that of Freuder and Mackworth.

Now consider Lines 15 to 27. A global consideration of the counters shows that if they never go negative, then there are at most:

$$\sum_{i=1}^{em} m \quad = \quad em^2$$

decrementations. Another way to see this is to consider the bounds on the two loops at Lines 15 and 18. First, we remark that a label is eliminated at most once from an object because the table $M$ "remembers" that fact. Now, given that label $\lambda'$ has been eliminated from node $j$, the only labels that can be affected are those at nodes $i$ which have an edge to $j$. Let $d_j$ be the vertex degree at node $j$; then since $j$ can appear at most $a$ times at Line 17, and since there are at most $d_j m$ elements of $S_{j\lambda'}$ for given $j$, we have that Line 20 can be executed at most:

$$\sum_{j=1}^{n} md_j \quad = \quad m^2 \sum_{j=1}^{n} d_j = m^2 e$$

The lower bound for arc consistency algorithms is easily found by assuming that the network is already arc consistent and seeing how much it costs to check. This requires evaluating $m^2$ pairs of labels at $e$ arcs. That is, any arc consistency algorithm has lower bound time complexity $\Omega(m^2 e)$, which in general is $\Omega(m^2 n^2)$. Since the lower bound time complexity for arc consistency is $\Omega(em^2)$ and the upper bound time complexity for AC-4 is $O(em^2)$, we have an optimal algorithm. We have already shown that the space required is on the same order as that required to define the relations.

We do not claim that there are no faster algorithms; the one we suggest here can be obviously improved: in step 1 we can remove from $A_i$ each $\lambda$ for which we have found that there is no more consistent labeling; this reduces the size of the $S_{j,\lambda}$ and therefore reduces the complexity of steps 1 and 2. However this is not a major improvement for the worst case. For planar graphs $e$ is of $O(n)$. AC-4 will run in $O(nm^2)$ and AC-3 in $O(nm^3)$ and both are linear in the number of nodes.

**Correctness of AC-4**

We outline here the key steps for a complete proof of the correctness of AC-4. The same approach can be used to prove AC-3.

1. Each label deleted from $A_i$ is not admissible for arc consistency.

2. Algorithm AC-1 is correct.

3. Let us suppose that AC-1 has already removed the labeling $\{(i_p, \lambda_p) \mid p = 1, \ldots, m\}$ and that AC-4 also removes them; let $(i, b)$ be the next labeling removed by AC-1. Then prove that $(i, \lambda)$ will also be removed by AC-4. Then by induction AC-4 removes at least as many labelings as AC-1.

**Space Complexity of AC-4**

The sets $S_{i\lambda}$ can be represented as linked lists and therefore use a space proportional to their number of elements: $O(em^2)$. The table $M$ has to be represented by an array of bits and therefore its size is $O(nm)$. We have at most $O(em)$ counters. Therefore, the total space required is $O(em^2)$. On real problems our algorithm never reaches its upper bound in space.

It should be noted that each algorithm must represent the graph and the possible labels for each of its nodes. This leads us to a minimal space requirement bounded by $O(e + nm)$. Algorithm AC-3 needs exactly this minimum upper bound.

Another approach is based on the observation that we can update the consistency of each of the label sets, $L_i$, in parallel. Recall that the labels in $L_i$ are consistent with their neighboring nodes if for each object $u_i$, each label in $L_i$ satisfies the constraint with at least one label in every other object's set of possible labels. Rosenfeld shows that there exists $L^\infty$ such that $L^\infty$ is consistent and for any consistent labeling $L', L' \subseteq L^\infty$. $L^\infty$ may be null; it may also be the case that no assignment of single labels to objects is consistent. This would mean that every binary constraint could be satisfied by pairwise choices of the labels, but that the choice of a single label for each object would leave some constraint unsatisfied; i.e., there would exist $u_i$ and $u_j$ such that $(\lambda_i, \lambda_j) \notin L_{ij}$, where $\lambda_i$ and $\lambda_j$ are the labels chosen for $u_i$ and $u_j$, respectively.

A parallel algorithm (called discrete relaxation) exists for finding $L^\infty$:

1. The initial labeling $L^0 = \{L_1, \ldots, L_n\}$, where $L_i = \Lambda_i$.

2. The $(k+1)$th labeling $L^{k+1}$ is obtained from the $k$th labeling $L^k$ by eliminating from each $L_i^k$ any label $\lambda$ such that $(\{\lambda\} \times L_j^k) * L_{ij} = 0$ for some $j$.

3. The algorithm halts at $L^\infty$ when no labels are discarded.

4. Since there is a finite number of labels, and labels are discarded at every step, there exists a finite $k$ such that $L^k = L^\infty$.

Once $L^\infty$ has been found, unambiguous labelings of the objects can be found. A depth-first tree search can be used, only now over a much reduced space, or search can be reduced even more by choosing a single label for some object and running the algorithm again and so on until a unique label is assigned for every object. The algorithm presented in Chapter 4 also computes arc consistency, and like Rosenfeld's algorithm is very useful when it can be performed in parallel or on fast hardware.

Discrete relaxation is an iterative parallel version of Waltz's filtering algorithm. Waltz designed this algorithm for disambiguation of the types of edges in line drawings of blocks-world scenes with shadows. that is, the objects are line segments in a line drawing representing a scene of polyhedra. Several interpretations are possible for each line segment, e.g., (1) a convex dihedral angle (both faces visible), (2) a concave dihedral angle (both faces visible), or (3) a dihedral angle (with one face visible). Initially, all of these must be considered possible for any given line segment. However, not all combinations of interpretatons are possible at a vertex, and these restrictions can be used to reduce the set of possible interpretations. Others who have investigated this problem domain include Clowes[6] and Huffman[42].

## 5.5 Path Consistency

Since it is too costly to solve the central problem, Montanari proposes that a path consistent network be found which is equivalent to the given network. Recall that a network is path consistent if and only if any pair of labels allowed by $R_{ij}$ is allowed by all paths from $u_i$ to $u_j$. That is, a pair is allowed if and only if at every intermediate $u_k$ a label is found that satisfies all unary and binary relations. One of the most interesting properties of a path consistent network is summarized in the following theorem.

**Theorem 5.5** *If every path of length 2 of a network with a complete graph is path consistent, then the network is path consistent.*

Based on this idea, Montanari proposed Algorithm PC-1 for computing the path consistent network for the network $N$ derived from the relation $R$.

```
procedure PC-1;
begin
  Y^n := R;
  repeat
    Y^0 := Y^n;
    for k := 1,n do
      for i := 1,n do
        for j := 1,n do
          Y^k_ij := Y^(k-1)_ij + Y^(k-1)_ij o Y^(k-1)_ij o Y^(k-1)_ij;
  until Y^n = Y^0;
  Y := Y^n;
end;
```

**Algorithm PC-1**. Montanari's path consistency algorithm.

Montanari proves that the result of PC-1 is the closure of the relation $R$. First, we need some more definitions. Let $b = (u_i, \lambda, u_j, \lambda')$ be an ordered pair of object-label pairs, where $\lambda \in \Lambda_i$ and $\lambda' \in \Lambda_j$.

**Definition 5.2** *A* path *is defined as* $p = (u_i = u_{i_0}, \ldots, u_{i_k} = u_j)$ *for* $k > 0$, *in the complete network* $N'$ *from* $u_i$ *to* $u_j$, *where each element of the path is an element of* $U$.

**Definition 5.3** *We say that* $b$ *is* allowed by p *if there exist labels* $\lambda = \lambda_0, \ldots, \lambda_k = \lambda'$ *such that* $\lambda_k \in \Lambda_k$, *and every pair of consecutive values satisfies the relation between the sets from which the values were chosen.*

**Definition 5.4** *An ordered pair* $b$ *is* legal *if allowed by all paths from* $u_i$ *to* $u_j$.

**Definition 5.5** *A network is* closed *if every* $b$ *is legal.*

Note that, although the nodes of the network graph are the $\Lambda_i$ sets, these correspond to the possible labels for the objects $u_i$. Thus, we can view the nodes as either the variables or their possible label sets.

**Theorem 5.6** *Algorithm PC-1 computes the closure* $Y$ *of* $N$. *In particular, if* $\mathbf{Y}^n_{ij}(r, s) = 1$ *in the network* $Y^n$, *then* $(u_i, \lambda, u_j, \lambda')$ *is allowed by all paths from* $u_i$ *to* $u_j$ *in* $N$.

**Proof:**

We prove that (i) $Y \subseteq N$ and $Y \equiv N$; (ii) $Y$ is closed; (iii) $\forall$ closed network $Y' \ni Y' \equiv N$ and $Y' \subseteq N$, it is the case that $Y' \subseteq Y \subseteq N$.

(i) First note that $Y^k \equiv Y^{k-1}$. It is clear that $Y^k \subseteq Y^{k-1}$. Conversly, suppose that for the following term in Line 9 of Algorithm PC-1 $\exists$ an $r, s \ni$:

$$\sum_{t=1}^{N_k} \mathbf{Y}_{ik}^{k-1}(r,t) * \mathbf{Y}_{kk}^{k-1}(t,t) * \mathbf{Y}_{kj}^{k-1}(t,s) = 0$$

Then there does not exist $\lambda \ni Y_{ik}^{k-1} = Y_{kk}^{k-1} = Y_{kj}^{k-1} = 1$. This means that $Y^k$ does not exclude any $n$-tuples not excluded by $Y^{k-1}$. Therefore, we have from the transitivity of inclusion and equivalence, $Y \equiv N$ and $Y \subseteq N$.

(ii) We show that $(u_i, \lambda, u_j, \lambda')$ allowed by $Y^n$ implies $(u_i, \lambda, u_j, \lambda')$ is legal in $Y^0$. Our induction assumption is that after Line 9 has been executed $(k-1)$ times, if $(u_i, \lambda, u_j, \lambda')$ is allowed by $Y_{ij}^{k-1}$, then it is legal in $Y^0$ for all paths with extrema in $u_i$ and $u_j$ and having $k-1$ as the maximum index for all intermediate vertexes. For $k = 1$, the assumption is obviously true. We now show that the assumption holds after the $k$th execution of Line 9. Let us consider any path $p$ from $u_i$ to $u_j$ having $k$ as the maximum index intermediate vertex. If $u_k$ is not on the path, then the induction step is proved. If it is on path $p$, then we can decompose $p$ into three subpaths:

1. A path from $u_i$ to $u_k$
2. A finite (maybe zero) number of cycles from $u_k$ to $u_k$
3. A path from $u_k$ to $u_j$

All those paths have $k-1$ as a maximum intermediate vertex index. In Line 9 we see that if $\mathbf{Y}_{ij}^{k}(r,s) = 1$, then a value $\lambda''$ can be found such that $\mathbf{Y}_{ik}^{k-1}(r,t) = 1, \mathbf{Y}_{kk}^{k-1}(t,t) = 1$, and $\mathbf{Y}_{kj}^{k-1}(t,s) = 1$. Therfore, by the induction assumption, suitable labels can be found for all the intermediate variables on paths (1), (2), and (3) such that the corresponding constraints in $Y^0$ are satisfied. When Algorithm PC-1 stops, we have $Y^n = Y^0 = Y$, and therefore $Y$ is closed.

(iii) Let $Y' \subseteq N$ be any closed network $\ni Y' \equiv N$. We must prove that in Line 9, $Y' \subseteq Y^k \forall k$. Then $Y' \subseteq Y$ will hold. Our induction hypothesis is then that $Y' \subseteq Y^{k-1}$ before the execution of Line 9. This is clearly true for the first execution of Line 9 in the first iteration $Y^0 = N$. Then we show that $Y' \subseteq Y^k$. That network $Y'$ is closed implies that:

$$Y'_{ij} \subseteq Y'_{ik} \circ Y'_{kj} + d_{ij} \text{ and } Y'_{kj} \subseteq Y'_{kk} \circ Y'_{kj}$$

Since the closure $Y$ of a network $N$ can be characterized by:

$$Y_{ij} = \sum_{k=1}^{n} R_{ik} Y_{kj} + d_{ij}$$

where $d_{ij} = \mathbf{I}_{ij}$ if $i = j$ and $d_{ij} = \mathbf{1}_{ij}$ otherwise ($\mathbf{I}$ is the identity and $\mathbf{1}$ is all 1's). Therefore, using the monotonicity of intersection and composition:

$$Y'_{ij} \subseteq Y'_{ik} \circ Y'_{kk} \circ Y'_{kj} + d_{ij}$$

and therefore:

$$Y'_{ij} = Y'_{ij} + Y'_{ik} \circ Y'_{kk} \circ Y'_{kj} + d_{ij}$$

But from our induction assumption and the monotonicity property we have:

$$Y'_{ik} \circ Y'_{kk} \circ Y'_{kj} + d_{ij} \quad \subseteq \quad Y^{k-1}_{ik} \circ Y^{k-1}_{kk} \circ Y^{k-1}_{kj} + d_{ij}$$

But the term $d_{ij}$ is redundant since for $i \neq j$, then $d_{ij} = 1$ and for $i = j$, then $d_{ij} = \mathbf{I}$. Thus, we have:

$$Y'_{ij} \subseteq Y^{k}_{ij} \qquad \textbf{Q.E.D.}$$

We have given Montanari's original proof here. For a shorter proof, see Mackworth [52]. Mackworth also points out that Algorithm PC-1 is a generalization of Algorithm AC-1. Moreover, he gives a more efficient algorithm: Algorithm PC-2.

```
procedure PC-2;
begin
   Q := {(i, k, j) | i ≤ j and ¬ (i = j = k)};
   while Q ≠ ∅ do
      begin
      remove (i, k, j) from Q;
      if revise(i, k, j) then Q := Q ⋃ Related_Paths(i, k, j);
      end;
end;

procedure revise(i, k, j);
begin
   Z := Y_ij * Y_ik ∘ Y_kk ∘ Y_kj;
   if Z = Y_ij then return false
      else Y_ij := Z; return true;
end;

procedure Related_Paths(i, k, j);
begin
   S_a := {(i, j, m) | i ≤ m ≤ n and m ≠ j}
          ⋃{(m, i, j) | 1 ≤ m ≤ j and m ≠ i}
          ⋃{(j, i, m) | j < m ≤ n}
          ⋃{(m, j, i) | 1 ≤ m < i};
   S_b := {(p, i, m) | 1 ≤ p ≤ m and 1 ≤ m ≤ n and ¬(p = i = m)
          and ¬(p = m = k)};
   if i < j then return S_a
      else return S_b.
end;
```

**Algorithm PC-2.** Mackworth's path consistency algorithm.

The same idea that we used to get Algorithm AC-2 applies for path consistency: when a pair of labelings is removed, Algorithm PC-2 considers only the length-2 paths that are related to the nodes of this pair. Algorithm PC-2 is of complexity $O(a^5n^3)$ whereas Algorithm PC-1 is $O(a^5n^5)$.

Due to Theorem 5 we know that full path consistency is equivalent to path consistency for all length-2 paths. Therefore, Algorithm PC-1 examines only these short paths. If a graph is not complete it can be completed by adding edges with the always *true* relation.

Just as we obtained Algorithm AC-4, a similar approach can be used to find a lower complexity path consistency algorithm; we give Han and Lee's version here[29] (see Algorithm PC-4).

**procedure** PC-4;

Step 1

```
M := 0; S_ibjc := ∅; List := ∅; Counter := 0;
for (i, j) ∈ E do
    for k = 1, n do
        for b ∈ A_i do
            for c ∈ A_j ∋ R_ij(b, c) = 1 do begin
                Total := 0;
                for d ∈ A_k do
                    if R_ik(b, d) = 1 and R_kj(d, c) = 1 then begin
                        Total := 0;
                        Append(S_ibkd, (j, c));
                        Append(S_jckd, (i, b));
                        end;
                    if Total = 0 then begin
                        M[i, b, j, c] := 1; M[j, c, i, b] := 1;
                        R_ij(b, c) := 0; R_ji(c, b) := 0;
                        end else begin
                        Counter[(i, b, j, c), k] := Total;
                        Counter[(j, c, i, b), k] := Total;
                        end;
                    end;
initialize List with {(i, b, j, c) | M[i, b, j, c] = M[j, c, i, b] = 1 and i < j};
```

Step 2

```
while List ≠ ∅ do begin
    choose and remove (k,d,l,e) from List;
    for (j, c) ∈ S_kdle do begin
        Counter[(k, d, j, c), l] := Counter[(k, d, j, c), l] − 1;
        Counter[(j, c, k, d), l] := Counter[(j, c, k, d), l] − 1;
        S_kdle := S_kdle − {(j, c)};
        S_jcle := S_jcle − {(k, d)};
        if Counter[(k, d, j, c), l] = 0 then
            if M[k, d, j, c] = 0 then begin
                M[k, d, j, c] := 1; M[j, c, k, d] := 1;
                Append(List, (k, d, j, c));
                R_kj(d, c) := 0; R_jk(c, d) := 0;
                end;
        end;
    for (j, c) ∈ S_lekd do begin
        Counter[(l, e, j, c), k] := Counter[(l, e, j, c), k] − 1;
        Counter[(j, c, l, e), k] := Counter[(j, c, l, e), k] − 1;
        S_lekd := S_lekd − {(j, c)};
        S_jckd := S_jckd − {(l, e)};
```

```
        if Counter[(l, e, j, c), k] = 0 then
            if M[l, e, j, c] = 0 then begin
                M[l, e, j, c] := 1; M[j, c, l, e] := 1;
                Append(List, (l, e, j, c));
                R_lj(e, c) := 0; R_jl(c, e) := 0;
                end;
    end;
end
```

**Algorithm PC-4.** Reduced complexity path consistency.

Clearly, the initialization step runs in $O(n^3m^3)$. The preparation of the List takes $O(n^3m^2)$. The pruning step removes an element of $R$ only once, and each such removal leads to at most $nm$ consistency checks. Therefore, the complexity of this part of the algorithm is $O(n^3m^3)$.

The space complexity is however very large: the number of counters is

$$\sum_{(i,j)\in U\times U} \mid A_i \mid \cdot \mid A_j \mid \cdot \mid \{k \in N \ni k \neq i, k \neq j\} \mid < n^3m^2$$

The space complexity of the whole algorithm is $O(n^3m^3)$. Because step 2 runs exactly in $O(n^3m^3)$ for a consistent network, this algorithm is truly cubic in its behavior.

Some optimization in space and time can be achieved. First, as was already mentioned in Reference 52, when exploring the length-2 paths, we can limit ourselves to the paths $(i, k, j)$ with $i < j$. This divides space and time by 2. Secondly, in step 2 we can update $\Lambda_i$ and $\Lambda_j$ each time one of $(i, \lambda)$ or $(j, \lambda')$ is put into *List*.

Algorithm AC-4 is an optimal algorithm for arc consistency. Algorithm PC-4 reduces the complexity of path consistency, but it still remains open whether Algorithm PC-4 is optimal. It is not obvious that any path consistency algorithm has to examine for each triple of nodes all possible labels in the worst case; if the answer is yes, then Algorithm PC-4 is optimal.

For practical cases, Algorithm AC-4 is easy to implement; however, it uses more space that Algorithm AC-3. Algorithm PC-4 is also easy to implement, however it may use a huge amount of space and therefore has to be run carefully. Using an algorithm like AC-4 in a backtrack search at each level of decision, it will run very fast on "common world" network problems providing the complete solution.

## 5.6 The Labeling Problem with Consistent Networks

The procedures to compute the path consistent network and the arc consistent network can be incorporated into search procedures to compute the set of globally consistent labels. Gaschnig[21] has described a general constraint satisfaction method and has shown how it can be used to solve certain classes of state space search problems. For sufficiently constrained problems, the method can eliminate locally inconsistent assignments and avoid search redundancy. Given a cryptarithmetic problem, e.g., DONALD + GERALD = ROBERT, the solution involves assigning a unique digit to each of the letters in the problem. Constraint relations can be

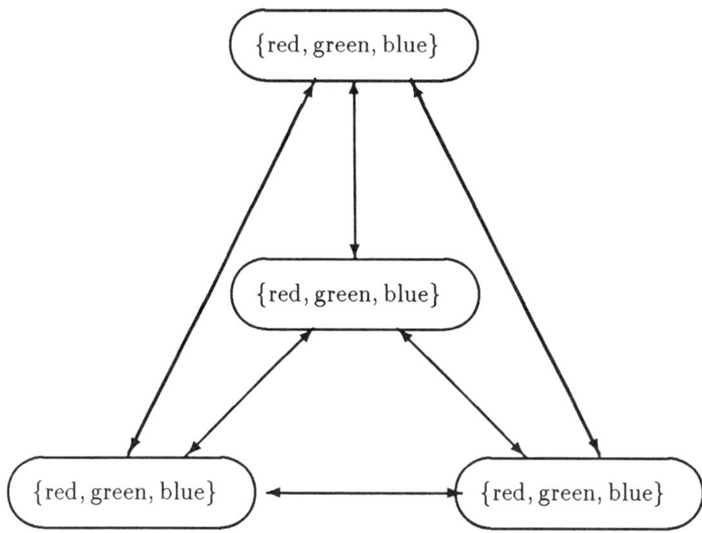


Figure 5.9: Graph to be colored.

defined between adjacent columns, and each letter can have only one digit associated with it. Gaschnig applied the algorithm to several other problems, including a two-dimensional packing problem, the Soma cube puzzle, and a space planning problem. Variations of discrete relaxation have been proposed by Ullmann[85] for finding subgraph isomorphisms and for pattern association, and by Hinton[39] for finding two-dimensional puppet figures.

The notions of arc consistency and path consistency can be generalized to a concept of $k$-consistency for any $k \leq n$, where $n$-consistency corresponds to global consistency[17]. The given constraints are represented by nodes in a constraint network. For all $k \leq n$, nodes representing $k$-ary constraints can be added to the constraint network. These constraints can be propagated in the augmented network to obtain higher-level consistency. The final result is a globally consistent network, where the $n$-ary node represents the $n$-ary constraint.

As an example of this idea, consider coloring the complete graph on four nodes with three colors (see Figure 5.9). Each node has as possible colors red, blue and green, and the arc between two nodes corresponds to the constraint: "not the same color as." This network is arc and path consistent, but there is no globally consistent set of labels. However, Freuder's algorithm reduces the set of possible labels to the empty set when the 4-ary relation is derived.

Another approach to finding consistent labelings as defined by an order $n$ relation is to define procedures or operators for reducing the size of the original relation[30]. The goal is to eliminate elements of the compatibility relation which do not contribute to any globally consistent labeling. Procedures are defined which accomplish this, but these can not be expected to reduce the worst case computational cost of solving the labeling problem because the labeling problem is NP-complete.

Let $S_r = \bigcap R_i \subseteq (A \times L)^N$ such that the $R_i$ have a common GCL; $S_r$ is called the *reduced relation* of $R$. The problem of reducing $R$ to $S_r$ is NP-complete. Haralick et al.[30] suggest two basic reduction operators. One way of reducing $R$ is to keep only those $2n$-tuples that are extendible to consistent labelings of $P$-tuples for some $P > n$. The operator can be defined as follows:

- $\Phi(P, R) = \{(a_1, \lambda_1, \ldots, a_N, \lambda_N) \in R \mid \forall a_{N+1}, \ldots, a_P \in A\ \exists \lambda_{N+1}, \ldots, \lambda_P \in L \ni (\lambda_1, \ldots, \lambda_P)$ is a consistent labeling of $(a_1, \ldots, a_P)\}$. Repeated application of $\Phi$ results in a relation $R'$ after a finite number of steps such that $\Phi(P, R') = R'$ and $S_r \subseteq R'$.

Another operator $\Psi(P, R)$ is defined by identifying object-label pairs that can be extended to consistent labelings and eliminate all $2N$-tuples that are not composed of such pairs. Let $T \subseteq A \times L$, $\Psi$ is then:

- $\Psi(T) = \{(a, \lambda) \in T \mid \forall a_1, \ldots, a_{P-1} \in A\ \exists \lambda_1, \ldots, \lambda_{P-1} \in L \ni (a_i, \lambda_i) \in T$, for $1 \leq i \leq p-1$, and $(\lambda_1, \lambda_2, \ldots, \lambda_{P-1}, \lambda)$is a consistent labeling of $(a_1, a_2, \ldots, a_{P-1}, a)\}$.

Let the projection of $R, \pi(R)$, be the set of $(a, \lambda) \in A \times L$ that are pairs of terms in $2N$-tuples in $R$. Then a finite number of applications of $\Psi$ to $T$ in $\pi(S_r)$ yields a relation $R'$ such that $\Psi(P, R') = R'$ and $S_r \subseteq R'$. Finally, if $R \subseteq (A \times L)^2$ and $\Phi(3, R) = R$, then $R$ is path consistent. If $R \subseteq (A \times L)^2$ is node consistent, then $R$ is arc consistent iff $\pi(R) = \Psi(2, \pi(R))$. These operators can be directly incorporated into a backtracking procedure for finding the GCLs of a relation $R$.

## 5.7 Performance Evaluation of Consistency Methods

It is hoped that using constraint reduction analysis will increase the speed and efficiency of problem solving. Claims have been made that network consistency algorithms are better than more standard procedures, e.g., pure backtracking. However, Gaschnig[21] presents experimental case studies comparing several algorithms, including network consistency algorithms and backtrack algorithms, and concludes that for the particular problem studied (the $N$ queens problem), certain backtracking algorithms were the most efficient in terms of the performance criteria chosen.

It should be noted that analytical evaluation of the techniques to be discussed is very difficult when only one solution is required. Results have been obtained for finding all the solutions[45], which is usually not a realistic approach. Therefore, it is useful to study the behavior of the algorithms on several standard problems.

As we have seen, consistency analysis is usually applied in conjunction with some standard search procedure, and in particular, with backtracking. At each variable instantiation, the discrete relaxation process is applied to the new network. In highly constrained problems, this can significantly reduce the possible labels at the remaining variables. However, the cost of applying the consistency algorithm is not negligible. It may even be the case that no significant reduction of the problem is achieved. Thus, in some problems, the overhead of the consistency analysis is simply added to the cost of the search, and in fact, slows down the search.

On the other hand, Mackworth[52] has conjectured that:

> ... the time taken to find a solution tends to be exponential in the number of variables, both in the worst case and on the average ...

and that Waltz-like algorithms

> ...are clearly more effective than automatic backtracking.

This led Gaschnig[21] to undertake a careful analysis of the performance of various consistency algorithms versus variations of backtracking. His results contradict Mackworth's conjecture. However, most of Gaschnig's results are based on problems with complete graphs, and Mackworth's conjecture may actually be true for sparse graphs.

Gaschnig compared several algorithms, including: backtrack, Waltz (the algorithm we call AC-2 and which Gaschnig calls "Domain Element Elimination Backtrack"), and three new algorithms proposed by Gaschnig which he called BACKMARK, BACKJUMP, and DEELEV. To better evaluate these algorithms, they are compared on a number of problems. These are separated into four categories:

1. (S1) The $N$-queens problem
2. (S2) Randomly permuted candidate values
3. (S3) Randomly generated SAP's matching N-queens
4. (S4) Randomly generated SAP's identical in size but varying in $l$, the degree of constraint

In order to perform an evaluation of the different algorithms, it is necessary to define some performance measures. The ones chosen by Gaschnig are:

- $T$: The number of pair tests
- $D$: The number of distinct pair tests
- $M$: The redundancy ratio ($M = T/D$)

These measures are produced during the search for one solution, not all solutions. For S2, S3, and S4, the measure is produced by taking the mean value of the performance measures produced over sample sets corresponding to each value of $N$.

The purpose of studying S2 is to compare the standard ordering of solution assignments to random orderings. S3 is important in order to generalize to broader sets of SAP's and to try and determine whether or not "families" of applications exist for which these techniques are preferable. The main purpose of these studies was to experimentally obtain some performance values. Then it is possible to test new theories against this set of quantitative data. The data was obtained from 13,000 distinct algorithm executions.

The major results can be summarized as follows:

1. BACKMARK performed best overall in that it had the least value for $T$ and $M$ appoaches closest to 1.
2. The Waltz algorithm executes more pair tests.
3. Randomizing the labels increases efficiency.
4. The Waltz algorithm does not do better on highly constrained problems.

These results are also supported by the random $N$-queens type of SAP's.

SAP's can be distinguished according to their graphs. A labeled directed graph whose $N$ nodes correspond with the $N$ problem variables and directed edges $e_{ij}$ and $e_{ji}$ connect $u_i$ and $u_j$ if and only if $P_{ij}$ is proper. If every $P_{ij}$ is proper, we have a complete graph and the SAP has a *complete consistency graph.* Otherwise, the graph is said to be incomplete. The $N$-queens problem has a complete graph; the edge labeling problem (usually) has an incomplete graph.

For example, when Waltz filtering is compared to BACKMARK on the $N$-queens problem, the total number of pair tests for Waltz filtering remains roughly an order of magnitude greater than for BACKMARK. At $N$=5 queens, $T$ is about 10 for BACKMARK and about 100 for Waltz filtering; at $N$=10 queens, $T$ is about 100 for BACKMARK and about 1,000 for Waltz filtering; at $N$=15 queens, $T$ is about 1,000 for BACKMARK and about 10,000 for Waltz filtering. This is typical of the results obtained across a wide spectrum of tests.

Further results have been obtained by Haralick and Elliot[32]. They have investigated exploiting two further principles:

1. First try places most likely to fail.

2. Remember what has been done: Avoid making the same mistake twice.

They proposed an algorithm called FORWARD CHECKING, and found that using the most likely to fail label performs better than standard backtracking, Ullman's, Waltz's, Mackworth's or Gaschnig's algorithms.

Their apporach is as follows. We say that a variable is a *past variable* which has already been assigned a label. The variable which is being assigned a label is the *current variable.* Any variables which have not yet been assigned labels and which are not yet under consideration are called *future variables.*

We can then see that lookahead algorithms such as Waltz filtering and discrete relaxation make sure that each future variable has at least one label compatible with past and current variables. Moreover, it is also the case that future variables' labels are compatible. The idea for forward checking is that not all future variables are checked with other future variables. In particular, there is only a partial lookahead into the future; namely, past and current and remembered are compared to the future variables and that is all.

The forward checking algorithm has an initial state in which it is assumed that no future variable has a label which is inconsistent with any past variable. Then the "remembers" function removes any inconsistent future variable label. If some future variable has no label in the variable label table, then the variable label table must be reset there, and the next current label is chosen. If there is no next current label, then backtrack to the previous variable.

It turns out that forward checking is similar to back checking except that back checking works only with past variables. Gaschnig's most powerful algorithm, BACKMARKing, works as follows. Let the current variable be $u$. Let $v$ be the lowest ordered variable backtracked to since the last visit to $u$. Remember $v$. If $v = u$, then perform backchecking. If $v < u$, then since all the labels for variable $u$ had been tested in the last visit to $u$, any label needing testing need only be tested against $v$ to $u - 1$ variables (where the labels may have changed).

The experiments performed indicate the following suggestions:

1. Don't use future variable versus future variable checking.

2. Use the node with the fewest labels first.

## 5.8 Other Constraint-Based Systems

Several approaches have been proposed for incorporating constraints directly into programming languages. One of the most notable is a constraint-oriented graphics language called "ThingLab" which is a constraint-oriented simulation laboratory developed by Borning[4]. His language is an extension of Smalltalk. The constraints specify relations that must hold between parts of a design or an experiment. The language is object-oriented in that there are objects representing structures and data types and there are well-defined operations on those structures. All nonprimitive objects are constructed hierarchically.

The constraints are defined as a rule and a set of methods for satisfying the constraints. The particular constraint satisfaction techniques used are to incrementally analyze constraint interactions and compile results of the analysis into executable code. This takes place in two passes. First a planning stage is performed which consists of a constraint satisfaction plan which is formulated and compiled. Then at run time, the appropriate compiled code is evoked.

Another constraint-based system is ALICE[51]. This is a general constraint satisfaction system which was designed to solve most combinatorial problems using a unified and general strategy. It is based upon backtracking with AC-4 applied at each node in the search. Finally, a more general system for describing and satisfying algebraic constraints (called Magritte) has been reported by Gosling[24]. He focuses on the problem of performing constraint satisfaction in an interactive layout system. Constraints are generally restricted to equations in conventional algebra.

There are two parts to Magritte:

1. Satisfaction: Given a constraint network, solve it.

2. Transformation: Given a constraint network, transform parts of the net into other forms.

Several advances are achieved by the system. The use of breadth first search to plan the propagation was found to be useful. The use of transformations to break loops seems to improve the performance of the system. Moreover, the automatic invocation of the transformation component from the constraint satisfaction component tends to relieve the user of having to deal with such very complicated and time-consuming issues. Finally, the use of a fast graph isomorphism algorithm makes the approach possible.

Another approach is to define classes of problems which map onto simple graph structures (e.g., trees). A good deal of work has been done in this area by Pearl and others[63,10,11,16]. It is often the case that tightly coupled heuristics can also be determined in such circumstances.

Thus, we have seen that there is a wide range of constraint techniques, and that their use is not restricted to just labeling problems. Indeed their use can be expanded into use for general programming systems. Moreover, there are many as yet unsolved problems in this area of research.

# Chapter 6

# Applications of Discrete Relaxation

We have studied the theory of discrete relaxation and have seen several approaches to solving the labeling problem. In this chapter we will examine in more detail several applications of discrete relaxation techniques. Given the wide range of applications, we cannot go into detail on them all, however, we will cover examples which are representative of the problems most commonly attacked with these methods.

Many problems can be posed in terms of discrete relaxation. For a good overview on the subject, see Haralick and Shapiro[31]. Applications include:

- Packing problems (including N-queens)
- Edge interpretation
- Satisfiability in the propositional calculus
- Latin square puzzles
- Graph homomorphisms
- Automata Homomorphisms
- Graph coloring
- Scene labeling

In this chapter, we describe in detail the application of discrete relaxation to the graph labeling problem and to the analysis of edges in a scene.

## 6.1 A Comparison of the AC Algorithms

Before considering specific application domains, let us compare the AC algorithms developed in Chapter 5. Consider the labeling problem shown in Figure 6.1. The relation $R$ is "lexicographically less than" and is indicated by the directed arcs. Even though the arcs are directed, they imply existence constraints in both directions; namely, for each label at the start of the arrow there must be a larger label at the

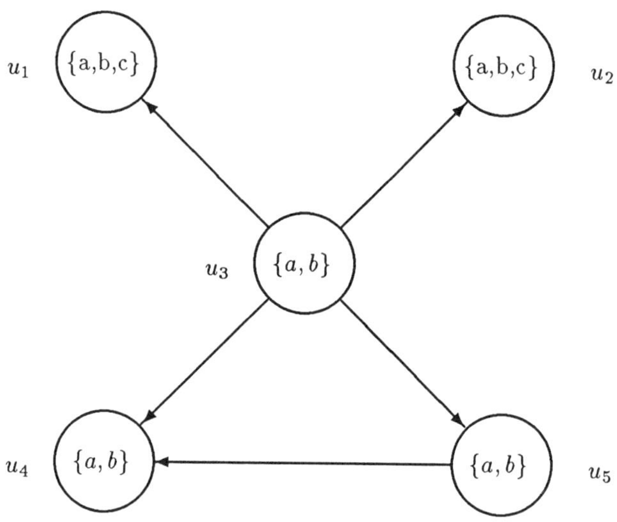


Figure 6.1: Example labeling problem.

end; while for a label at the end, there must be a smaller label at the start. Note that no assignment of labels satisfies the constraints.

Consider the following trace of AC-1:

| iteration # | (i,j) | Revise | $L_1$ | $L_2$ | $L_3$ | $L_4$ | $L_5$ |
|---|---|---|---|---|---|---|---|
| | | | $\{a, b, c\}$ | $\{a, b, c\}$ | $\{a, b\}$ | $\{a, b\}$ | $\{a, b\}$ |
| 1 | (3,1) | | | | | | |
| | (1,3) | $L_1 - \{a\}$ | $\{b, c\}$ | | | | |
| | (3,2) | | | | | | |
| | (2,3) | $L_2 - \{a\}$ | | $\{b, c\}$ | | | |
| | (3,5) | $L_3 - \{b\}$ | | | $\{a\}$ | | |
| | (5,3) | $L_5 - \{a\}$ | | | | | $\{b\}$ |
| | (3,4) | | | | | | |
| | (4,3) | $L_4 - \{a\}$ | | | | $\{b\}$ | |
| | (5,4) | $L_5 - \{b\}$ | | | | | $\emptyset$ |
| | (4,5) | $L_4 - \{b\}$ | | | | $\emptyset$ | |
| 2 | (3,1) | | | | | | |
| | (1,3) | $L_1 - \{c\}$ | $\{b\}$ | | | | |
| | (3,2) | | | | | | |
| | (2,3) | $L_2 - \{c\}$ | | $\{b\}$ | | | |
| | (3,5) | $L_3 - \{a\}$ | | | $\emptyset$ | | |
| | (5,3) | | | | | | |
| | (3,4) | | | | | | |
| | (4,3) | | | | | | |
| | (5,4) | | | | | | |

| | | | | |
|---|---|---|---|---|
| | (4,5) | | | |
| 3 | (3,1) | | | |
| | (1,3) | $L_1 - \{b\}$ | $\emptyset$ | |
| | (3,2) | | | |
| | (2,3) | $L_2 - \{b\}$ | | $\emptyset$ |
| | (3,5) | | | |
| | (5,3) | | | |
| | (3,4) | | | |
| | (4,3) | | | |
| | (5,4) | | | |
| | (4,5) | | | |
| 4 | ... | – no change | | |

This procedure could be improved by stopping once any domain is reduced to the empty set because there is no support for any label at other nodes in such a case. This causes all the label sets to go empty. However, if absolute labels are imposed (i.e., labels which are retained even without support), then the process must be allowed to converge.

Next consider a trace of AC-3:

| Arc | $Q$ | Revise | $L_1$ | $L_2$ | $L_3$ | $L_4$ | $L_5$ |
|---|---|---|---|---|---|---|---|
| | $\{(3,1)(1,3)(3,2)$ $(2,3)(3,5)(5,3)$ $(3,4)(4,3)(5,4)(4,5)\}$ | | $\{a,b,c\}$ | $\{a,b,c\}$ | $\{a,b\}$ | $\{a,b\}$ | $\{a,b\}$ |
| (3,1) | $\{(1,3)(3,2)(2,3)$ $(3,5)(5,3)(3,4)$ $(4,3)(5,4)(4,5)\}$ | | | | | | |
| (1,3) | $\{(3,2)(2,3)(3,5)$ $(5,3)(3,4)(4,3)$ $(5,4)(4,5)\}$ | $L_1 - \{a\}$ | $\{b,c\}$ | | | | |
| (3,2) | $\{(2,3)(3,5)(5,3)$ $(3,4)(4,3)(5,4)(4,5)\}$ | | | | | | |
| (2,3) | $\{(3,5)(5,3)(3,4)$ $(4,3)(5,4)(4,5)\}$ | $L_2 - \{a\}$ | | $\{b,c\}$ | | | |
| (3,5) | $\{(2,3)(1,3)(5,3)$ $(3,4)(4,3)(5,4)(4,5)\}$ | $L_3 - \{b\}$ | | | $\{a\}$ | | |
| (2,3) | $\{(1,3)(5,3)(3,4)$ $(4,3)(5,4)(4,5)\}$ | $L_2 - \{c\}$ | | $\{b\}$ | | | |
| (1,3) | $\{(5,3)(3,4)(4,3)$ $(5,4)(4,5)\}$ | $L_1 - \{c\}$ | $\{b\}$ | | | | |
| (5,3) | $\{(3,4)(4,3)(5,4)(4,5)\}$ | $L_5 - \{a\}$ | | | | | $\{b\}$ |
| (3,4) | $\{(4,3)(5,4)(4,5)\}$ | | | | | | |
| (4,3) | $\{(5,4)(4,5)\}$ | $L_4 - \{a\}$ | | | | $\{b\}$ | |
| (5,4) | $\{(3,5)(4,5)\}$ | $L_5 - \{b\}$ | | | | | $\emptyset$ |
| (3,5) | $\{(4,3)(2,3)(1,3)(4,5)\}$ | $L_3 - \{a\}$ | | | $\emptyset$ | | |
| (4,3) | $\{(5,4)(2,3)(1,3)(4,5)\}$ | $L_4 - \{b\}$ | | | | $\emptyset$ | |
| (5,4) | $\{(2,3)(1,3)(4,5)\}$ | | | | | | |

| | | | |
|---|---|---|---|
| (2,3) | $\{(1,3)(4,5)\}$ | $L_2 - \{b\}$ | $\emptyset$ |
| (1,3) | $\{(4,5)\}$ | $L_1 - \{b\}$ | $\emptyset$ |
| (4,5) | $\emptyset$ | | |

Notice that arcs are added to the queue when arcs (3,5), (4,3), (5,4), (3,5), and (4,3) are examined.

Finally, to trace AC-4 requires a good deal more detail:

*Initialization Step*

| $(i,j)$ | $(\lambda,\lambda')$ | $R(\lambda,\lambda')$ | counter | $S_{j\lambda'}$ | $M_{i\lambda}$ |
|---|---|---|---|---|---|
| (3,1) | (a,a) | False | 0 | | |
| | (a,b) | True | 1 | $S_{1b} = \{(3,a)\}$ | |
| | (a,c) | True | 2 | $S_{1c} = \{(3,a)\}$ | |
| | | counter[(3,1),$a$] := 2 | | | $M_{3a} = 0$ |
| | (b,a) | False | 0 | | |
| | (b,b) | False | 0 | | |
| | (b,c) | True | 1 | $S_{1c} = \{(3,a)(3,b)\}$ | |
| | | counter[(3,1),$b$] := 1 | | | $M_{3b} = 0$ |
| (1,3) | (a,a) | False | 0 | | |
| | (a,b) | False | 0 | | |
| | | | | | $M_{1a} = 1$ |
| | (b,a) | True | 1 | $S_{3a} = \{(1,b)\}$ | |
| | (b,b) | False | 1 | | |
| | | counter[(1,3),$b$] := 1 | | | $M_{1b} = 0$ |
| | (c,a) | True | 1 | $S_{3a} = \{(1,b),(1,c)\}$ | |
| | (c,b) | True | 2 | | |
| | | counter[(1,3),$c$] := 2 | | | $M_{1c} = 0$ |
| (3,2) | (a,a) | False | 0 | | |
| | (a,b) | True | 1 | $S_{2b} = \{(3,a)\}$ | |
| | (a,c) | True | 2 | $S_{2c} = \{(3,a)\}$ | |
| | | counter[(3,2),$a$] := 2 | | | $M_{3a} = 0$ |
| | (b,a) | False | 0 | | |
| | (b,b) | False | 0 | | |
| | (b,c) | True | 1 | $S_{2c} = \{(3,a)(3,b)\}$ | |
| | | counter[(3,2),$b$] := 1 | | | $M_{3b} = 0$ |
| (2,3) | (a,a) | False | 0 | | |
| | (a,b) | False | 0 | | |
| | | | | | $M_{2a} = 1$ |
| | (b,a) | True | 1 | $S_{3a} = \{(1,b)(1,c)(2,b)\}$ | |
| | (b,b) | False | 1 | | |
| | | counter[(2,3),$b$] := 1 | | | $M_{2b} = 0$ |
| | (c,a) | True | 1 | $S_{3a} = \{(1,b)(1,c)(2,b)(2,c)\}$ | |
| | (c,b) | True | 2 | $S_{3b} = \{(1,c)(2,c)\}$ | |
| | | counter[(2,3),$c$] := 2 | | | $M_{2c} = 0$ |
| (3,5) | (a,a) | False | 0 | | |

| | | | | | |
|---|---|---|---|---|---|
| | (a,b) | True | 1 | $S_{5b} = \{(3,a)\}$ | |
| | | counter[(3,5),$a$] := 1 | | | $M_{3a} = 0$ |
| | (b,a) | False | 0 | | |
| | (b,b) | False | 0 | | |
| | | | | | $M_{3b} = 1$ |
| (5,3) | (a,a) | False | 0 | | |
| | (a,b) | False | 0 | | |
| | | | | | $M_{5a} = 1$ |
| | (b,a) | True | 1 | $S_{3a} = \{(1,b)(1,c)(2,b)(2,c)(5,b)\}$ | |
| | (b,b) | False | 1 | | |
| | | counter[(5,3),$b$] := 1 | | | $M_{5b} = 0$ |
| (3,4) | (a,a) | False | 0 | | |
| | (a,b) | True | 1 | $S_{4b} = \{(3,a)\}$ | |
| | | counter[(3,4),$a$] := 1 | | | $M_{3a} = 0$ |
| | (b,a) | False | 0 | | |
| | (b,b) | False | 0 | | |
| | | | | | $M_{3b} = 1$ |
| (4,3) | (a,a) | False | 0 | | |
| | (a,b) | False | 0 | | |
| | | | | | $M_{4a} = 1$ |
| | (b,a) | True | 1 | $S_{3a} = \{(1,b)(1,c)(2,b)(2,c)$ $(5,b)(4,b)\}$ | |
| | (b,b) | False | 1 | | |
| | | counter[(4,3),$b$] := 1 | | | $M_{4b} = 0$ |
| (5,4) | (a,a) | False | 0 | | |
| | (a,b) | True | 1 | $S_{4b} = \{(3,a)(5,a)\}$ | |
| | | counter[(5,4),$a$] := 1 | | | $M_{5a} = 1$ |
| | (b,a) | False | 0 | | |
| | (b,b) | False | 0 | | |
| | | | | | $M_{5b} = 1$ |
| (4,5) | (a,a) | False | 0 | | |
| | (a,b) | False | 0 | | |
| | | | | | $M_{4a} = 1$ |
| | (b,a) | True | 1 | $S_{5a} = \{(4,b)\}$ | |
| | (b,b) | False | 1 | | |
| | | counter[(4,5),$b$] := 1 | | | $M_{4b} = 0$ |

From this, we have the following:

$$List = \{(1,a)(2,a)(3,b)(4,a)(5,a)(5,b)\}$$

$$\begin{aligned} S_{5a} &= \{(4,b)\} \\ S_{4b} &= \{(3,a)(5,a)\} \\ S_{3a} &= \{(1,b)(1,c)(2,b)(2,c)(5,b)(4,b)\} \\ S_{5b} &= \{(3,a)\} \end{aligned}$$

$$
\begin{aligned}
S_{3b} &= \{(1,c)(2,c)\} \\
S_{2c} &= \{(3,a)(3,b)\} \\
S_{2b} &= \{(3,a)\} \\
S_{1c} &= \{(3,a)(3,b)\} \\
S_{1b} &= \{(3,a)\}
\end{aligned}
$$

$M$ is:

| | $a$ | $b$ | $c$ |
|---|---|---|---|
| 1 | 1 | 0 | 0 |
| 2 | 1 | 0 | 0 |
| 3 | 0 | 1 | – |
| 4 | 1 | 0 | – |
| 5 | 1 | 1 | – |

and counter is:

| | $a$ | $b$ | $c$ |
|---|---|---|---|
| (3,1) | 2 | 1 | – |
| (1,3) | 0 | 1 | 2 |
| (3,2) | 2 | 1 | – |
| (2,3) | 0 | 1 | 2 |
| (3,5) | 1 | 0 | |
| (5,3) | 0 | 1 | |
| (3,4) | 1 | 0 | |
| (4,3) | 0 | 1 | |
| (5,4) | 1 | 0 | |
| (4,5) | 0 | 1 | |

Notice that $M[5,a] = M[5,b] = 1$, which means that both the possible labels for node 5 will be removed, and therefore the algorithm could terminate at this point. Tracing step 2 of Algorithm AC-4:

*Step 2*

| $(j,\lambda')$ | $(i,\lambda)$ | counter | $M_{i\lambda}$ | List |
|---|---|---|---|---|
| | | | | $\{(1,a)(2,a)(3,b)(4,a)(5,a)(5,b)\}$ |
| $(1,a)$ | $\emptyset$ | | | $\{(2,a)(3,b)(4,a)(5,a)(5,b)\}$ |
| $(2,a)$ | $\emptyset$ | | | $\{(3,b)(4,a)(5,a)(5,b)\}$ |
| $(3,b)$ | $(2,c)$ | $c[(2,3),c] := 1$ | | $\{(4,a)(5,a)(5,b)\}$ |
| | $(1,c)$ | $c[(1,3),c] := 1$ | | $\{(4,a)(5,a)(5,b)\}$ |
| (4,a) | $\emptyset$ | | | $\{(5,a)(5,b)\}$ |
| (5,a) | $(4,b)$ | $c[(4,5),b] := 0$ | $M_{4b} := 1$ | $\{(4,b)(5,b)\}$ |
| (4,b) | $(3,a)$ | $c[(3,4),a] := 0$ | $M_{3a} := 1$ | $\{(3,a)(5,b)\}$ |
| | $(5,a)$ | $c[(5,4),a] := 0$ | $M_{5a} = 1$ | $\{(3,a)(5,b)\}$ |
| (3,a) | $(1,b)$ | $c[(1,3),b] := 0$ | $M_{1b} := 1$ | $\{(1,b)(5,b)\}$ |
| | $(1,c)$ | $c[(1,3),c] := 0$ | $M_{1c} := 1$ | $\{(1,c)(1,b)(5,b)\}$ |
| | $(2,b)$ | $c[(2,3),b] := 0$ | $M_{2b} := 1$ | $\{(2,b)(1,c)(1,b)(5,b)\}$ |
| | $(2,c)$ | $c[(2,3),c] := 0$ | $M_{2c} := 1$ | $\{(2,c)(2,b)(1,c)(1,b)(5,b)\}$ |
| | $(5,b)$ | $c[(5,3),b] := 0$ | $M_{5b} := 1$ | $\{(2,c)(2,b)(1,c)(1,b)(5,b)\}$ |
| | $(4,b)$ | $c[(4,3),b] := 0$ | $M_{4b} := 1$ | $\{(2,c)(2,b)(1,c)(1,b)(5,b)\}$ |

| | | | | |
|---|---|---|---|---|
| (2,c) | $(3,a)$ | $c[(3,2),a] := 1$ | | $\{(2,b)(1,c)(1,b)(5,b)\}$ |
| | $(3,b)$ | $c[(3,2),b] := 0$ | $M_{3b} := 1$ | $\{(2,b)(1,c)(1,b)(5,b)\}$ |
| (2,b) | $(3,a)$ | $c[(3,2),a] := 0$ | $M_{3a} := 1$ | $\{(1,c)(1,b)(5,b)\}$ |
| (1,c) | $(3,a)$ | $c[(3,1),a] := 1$ | | $\{(1,b)(5,b)\}$ |
| | $(3,b)$ | $c[(3,1),b] := 0$ | $M_{3b} := 1$ | $\{(1,b)(5,b)\}$ |
| (1,b) | $(3,a)$ | $c[(3,1),a] := 0$ | $M_{3a} := 1$ | $\{(5,b)\}$ |
| (5,b) | $(3,a)$ | $c[(3,5),a] := 0$ | $M_{3a} := 1$ | $\emptyset$ |

Looking at the $(j, \lambda')$ pairs, we see that all labels have been deleted, and that the empty labeling is the only consistent labeling.

These examples should offer some insight into the implementation and the mechanism of the AC algorithms. In Chapter 8, we give parallel versions of these algorithms.

## 6.2 Waltz Filtering

Given a set of edges from a digital image, an important problem is the interpretation of the edges in terms of their organization on polyhedra. Usually, we assume that we have an ideal line drawing of the edges laying on a set of polyhedra in a scene. The use of constraints in the analysis of such scenes began very early and was reported by Clowes[6], Guzman[28] and, of course, Waltz[88]. Later, when Rosenfeld et al.[67] introduced the notion of discrete relaxation, they gave examples for solving the edge interpretation problem.

Most often the assumption is made that the scene may only have trihedral objects in it. This means that any vertex is formed by the meeting of exactly three surfaces. Given a line drawing of such a scene, we would like to interpret the edges in the scene in terms of the role they play in forming distinct polyhedra. This requires labeling each edge with a physical interpretation. Consider for the moment scenes devoid of cracks and shadows and with polyhedra in general position. In this case, all edges are one of the following:

1. *Convex*(+): A line from a point on the surface on one side of the edge to a point on the surface on the other side of the edge passes through the polyhedron.
2. *Concave*(−): A line from a point on the surface on one side of the edge to a point on the surface on the other side of the edge passes through empty space.
3. *Occluded*(← or →): The edge is the result of one polyhedron occluding another. This gives rise to an oriented label which indicates that the occluding surface lies to the right of the oriented vector. Thus, any edge might possibly be labeled with an occluding label oriented in one of two directions.

The edge interpretation problem is then to assign the appropriate label from {+, −, ←, →} to each edge. We show an example of a polyhedral scene in Figure 6.2, and the corresponding labeled scene in Figure 6.3. Several authors have shown how to extend this analysis to wider domains of scenes including shadows, cracks, and nontrihedral objects[75,82,88].

Constraints on the possible combinations of edge labelings can be obtained from the relations that the edges must satisfy at the vertexes of trihedral objects. Guzman[28] analyzed the possible vertex types in such a scene and gave names to them. For example, an example of a T vertex occurring in Figure 6.3 is shown in Figure 6.4. The physical constraints of the way that the different kinds of edges

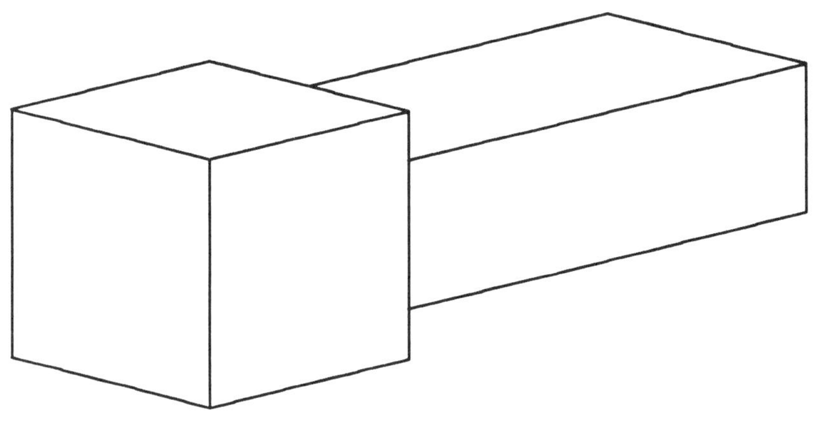

Figure 6.2: Polyhedral scene.

can come together at a vertex considerably limit the possible labelings of the edges in a scene. Figure 6.5 shows that only four labelings are possible for the T vertex, whereas, in general, there are sixty-four possibilities. Three of the four can be found in Figure 6.3.

Let us now set up the problem more formally. If $N$ edges meet at a vertex, we say that the vertex is an order $N$ vertex. For example, the T vertex is an order 3 vertex. An $N$th-order constraint relation is induced by the set of order $N$ vertexes in a scene. Let the set of $N$th-order vertexes be denoted by $V$, let the set of line segments (edges) be denoted by $U = \{1, ..., n\}$, let $\Lambda = \{+, -, \leftarrow, \rightarrow\}$ be the set of labels, and let $X = \{T, A, L, Y\}$ be the set of vertex types. The vertex types are shown in Figure 6.6.

The problem to be solved is to label each edge in the scene with an edge type (e.g., one of $\{+, -, \leftarrow, \rightarrow\}$) such that the total set of labels on the edges is consistent. That is, every vertex has a physically possible and consistent set of edge labels on each edge issuing from the vertex. In order to describe consistency, we must first give a convention for labeling the four vertex types. We will consider that each L vertex type will be labeled by a pair, and that each T, A, and Y type vertex will be labeled by a triple whose order is crucial; these orderings are given by Figure 6.7. We can then define the consistency relations imposed by the physical constraints. In order to write these as triples and pairs, we will further adopt the convention that:

$\leftarrow$ The arrow points away from the vertex.

$\rightarrow$ The arrow points towards the vertex.

With this convention, we can now give the relations both pictorially and below them in vector notation (see Figure 6.8). For more on the derivation of physical constraints, see Winston[92]. These relations specify $N$-tuples of labels that can

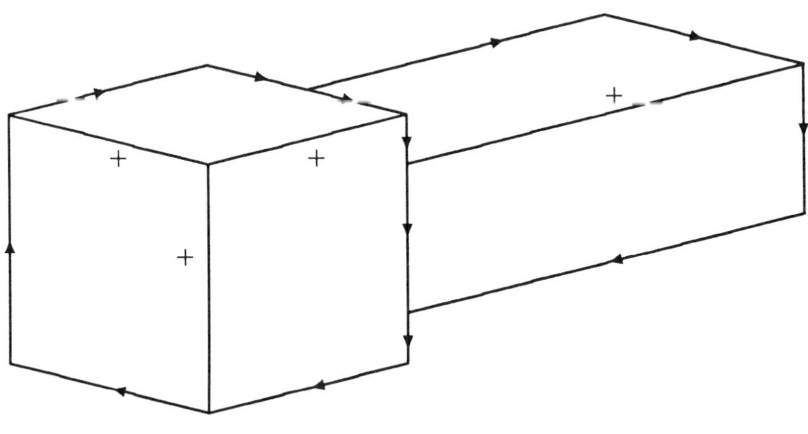


Figure 6.3: Edge labeling of polyhedral scene.

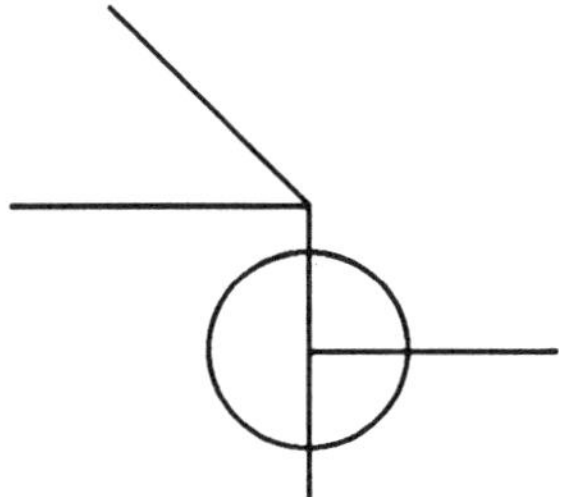

Figure 6.4: An example T vertex.

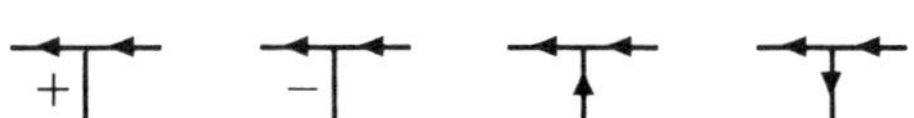


Figure 6.5: Possible labelings of T vertex.

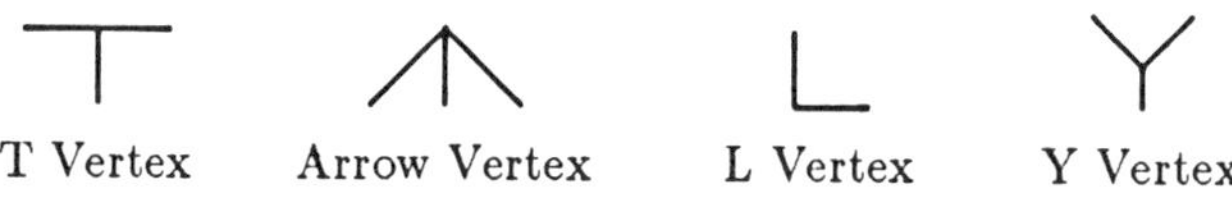


Figure 6.6: The four vertex types.

Figure 6.7: The four vertex edge orderings.

meet at each type of vertex.

To solve the labeling problem, we now need to formulate this as one of the consistency algorithms seen in the previous chapter. Taking the simplest discrete relaxation algorithm, we must perform the following steps until there is no change in the label sets:

$\forall$ edge in the edge set
  $\forall$ label still possible for that edge
    $\forall$ vertex type that that edge is a part of
      **if** the label is not permitted by the vertex consistency relation $R_x$
        **then** eliminate that label from the edge's possible label set.

Which relation $R_x$ to use is determined by the type $x$ of the vertex under consideration.

As a simple example, let's reconsider the scene in Figure 6.2. Suppose that we have already labeled the boundary edges by some simple boundary-following technique (see Figure 6.9). Note that the vertexes are labeled $v_1$ through $v_{13}$, and that the edges are named 1 through 17. The $\Lambda_i$'s of initial labels for the edges are given in Figure 6.10. We use $(i \rightarrow j)$ to indicate the direction of the arrow from vertex $v_i$ to $v_j$.

Our goal is to produce a consistent labeling of the edges. Using the algorithm described above to perform the discrete relaxation, we can trace some of the calculation. Suppose that we first look at edge 1. It only has one label, namely $(1 \rightarrow 2)$. Checking each vertex that it is a part of (i.e., $v_1$ and $v_2$), we see that $(\rightarrow, \leftarrow, +)$ satisfies $v_1 \in R_A$ and that $(\leftarrow, \rightarrow)$ satisfies $v_2 \in R_L$. Thus, $\Lambda_1$ is consistent. Upon checking edge 3, however, we must check all four possible labels. First, we check the label $(1 \rightarrow 11)$. In order for this label to be consistent with $v_1$, there must be some triple $(x, y, \rightarrow) \in R_A$. Since there is no such element in the relation, then the

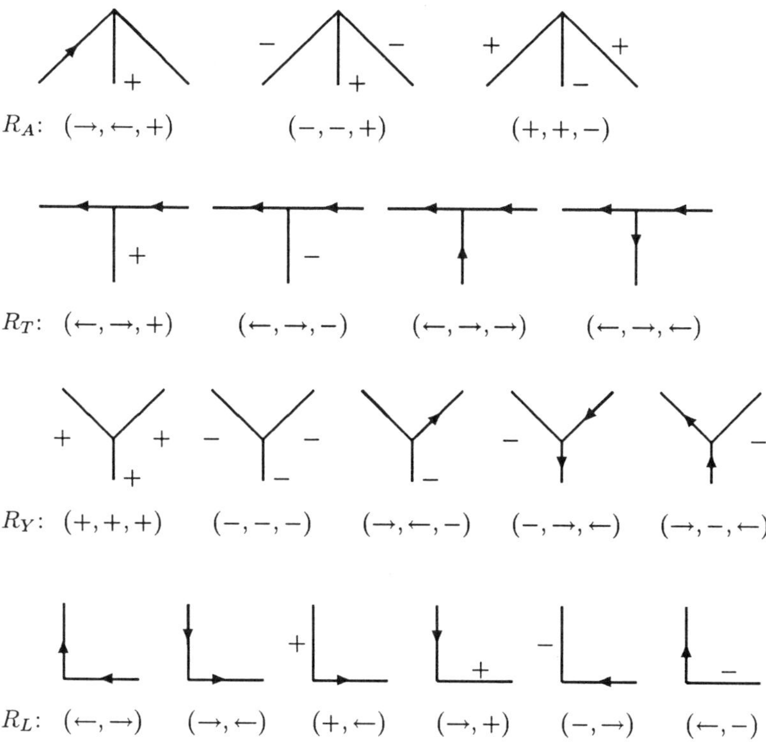


Figure 6.8: The consistent labelings of the four vertex types.

label → must be eliminated from $\Lambda_3$. Likewise, just from vertex $v_1$, we find that the labels ← and − must be eliminated. This leaves only + as a possible label. Continuing in this manner, we find the consistent labeling given in Figure 6.11. It is perhaps instructive to repeat this problem without giving the boundary edges in the initial labeling. In this case, the final consistent labeling is given in Figure 6.12. In this case it can be seen that the result is not unambiguous, and, in fact, without applying backtracking, it cannot be determined if there is indeed an unambiguous consistent labeling.

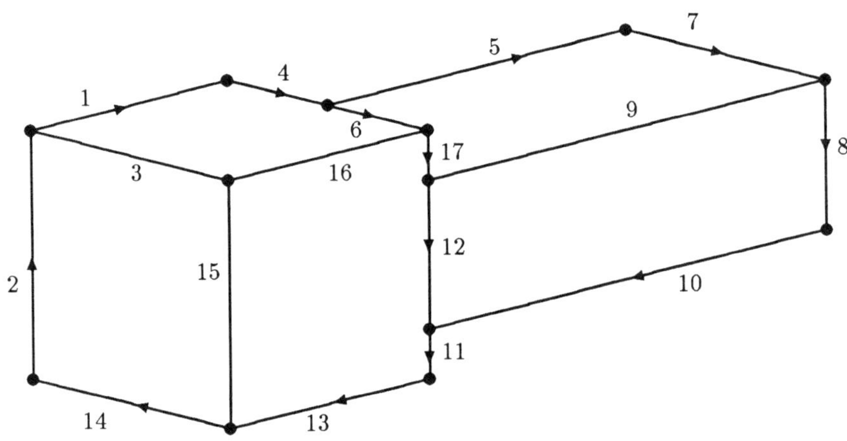


(a) Edge numbers

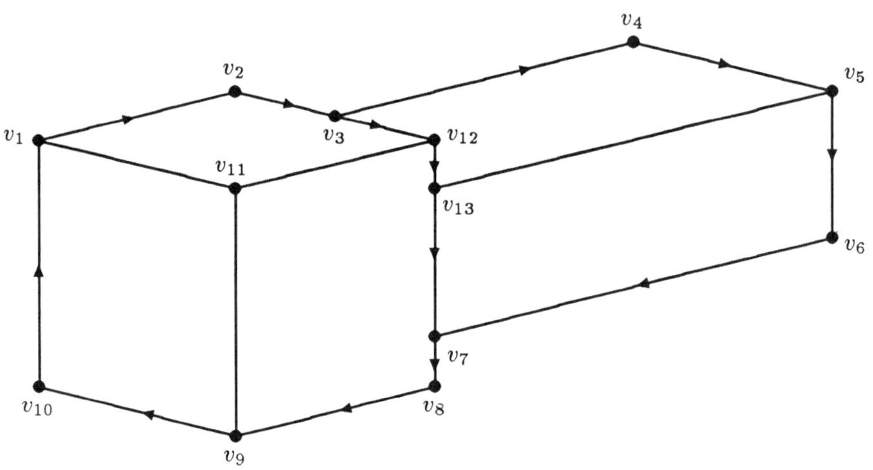


(b) Vertex numbers

Figure 6.9: Initial boundary labeled polyhedral scene.

$\Lambda_1 : (1 \rightarrow 2)$
$\Lambda_2 : (10 \rightarrow 1)$
$\Lambda_3 : (1 \rightarrow 11), (11 \rightarrow 1), +, -$
$\Lambda_4 : (2 \rightarrow 3)$
$\Lambda_5 : (3 \rightarrow 4)$
$\Lambda_6 : (3 \rightarrow 12), (12 \rightarrow 3), +, -$
$\Lambda_7 : (4 \rightarrow 5)$
$\Lambda_8 : (5 \rightarrow 6)$
$\Lambda_9 : (5 \rightarrow 13), (13 \rightarrow 5), +, -$
$\Lambda_{10} : (6 \rightarrow 7)$
$\Lambda_{11} : (7 \rightarrow 8)$
$\Lambda_{12} : (7 \rightarrow 13), (13 \rightarrow 7), +, -$
$\Lambda_{13} : (8 \rightarrow 9)$
$\Lambda_{14} : (9 \rightarrow 10)$
$\Lambda_{15} : (9 \rightarrow 11), (11 \rightarrow 9), +, -$
$\Lambda_{16} : (11 \rightarrow 12), (12 \rightarrow 11), +, -$
$\Lambda_{16} : (12 \rightarrow 13), (13 \rightarrow 11), +, -$

Figure 6.10: Initial edge labels.

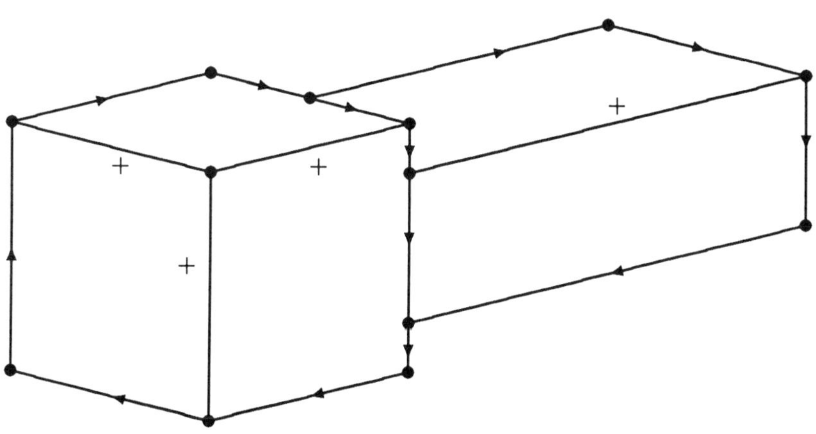


Figure 6.11: Unambiguous consistent labeling.

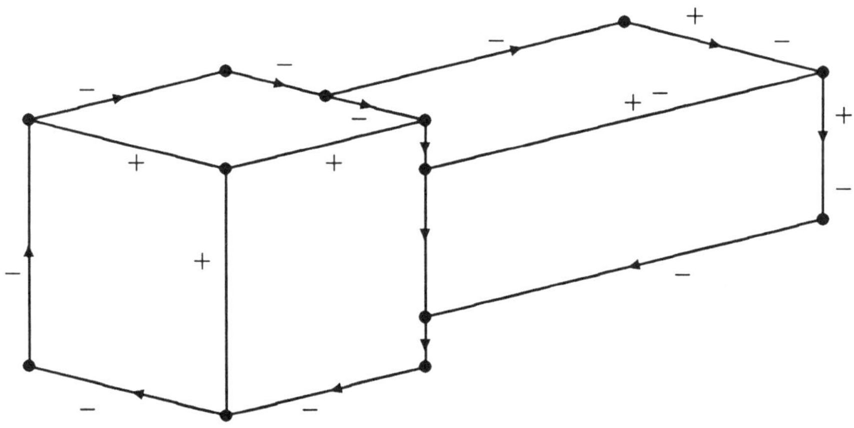

Figure 6.12: Ambiguous consistent labeling.

# Chapter 7

# Hierarchical Constraint Processes

This chapter presents an approach to shape representation and recognition based on a union of constraint propagation procedures and syntactic pattern analysis procedures due to Davis and Henderson[8,34,36].[1] Constraint propagation procedures have been applied to a wide variety of problems in computer vision, especially low-level vision. They can, to a great degree, overcome the intrinsic ambiguity to assigning descriptions (such as "horizontal edge," "vertical edge") to image parts (e.g., pixels). They operate by initially assigning a set of labels (descriptions) to each picture part and then iteratively updating each labeling based on labelings at "adjacent" parts and a model of label compatibility (usually expressed as a relation). Their advantages include locality, since only "adjacent" parts interact, and speed, since each iteration is performed in parallel to each picture part. A survey of the application of such techniques can be found in Reference 9.

Syntactic pattern analysis techniques have been extensively applied to problems in planar shape recognition and analysis[18,23]. In order to apply the syntactic paradigm, a shape must be segmented into pieces which we shall call the *primitives.* Each primitive must then be assigned a terminal symbol name from the grammar at hand. The resulting string (or cycle) can then be parsed by an appropriate parsing mechanism, which can even be designed to account for a limited number of missing symbols, extraneous symbols, and incorrect symbols. The advantages of the syntactic approach include its flexibility in analyzing noisy and distorted data and the organization it imposes on the data by generating a parse tree. Although it is possible in applying syntactic techniques to a recognition problem to design a specific grammar for each recognition class (such as a Boeing 747 or a DC 10), the more general approach, and the one adopted here, is to design a single grammar for the largest class of shapes (e.g., airplanes) and then to recognize specific shapes based on an analysis of the resulting parse trees (the engine position, for example, can be used to distinguish between a Boeing 747 and a DC 10). This leads to much larger grammars and often less efficient parsers, especially when the number of recognition classes is small.

There are, however, several shortcomings of this syntactic approach (some of which are partially overcome by top-down parsing; see Stockman[81]). They include:

[1] This chapter is a modified version of Reference 36.

1. The necessity for computing a unique *segmentation* of the shape primitives
2. The requirement of assigning a *single* terminal symbol name to each primitive

In computer vision it has been recognized that any single segmentation based on low-level analysis is bound to contain errors, i.e., missing segments, extraneous segments, etc., because of the many scales of events in an image[54]. One way to overcome the problem of missing segments is to allow for multiple, overlapping primitives on parts of the shape which are difficult to segment.

Furthermore, it is also the case that a premature commitment to a single description of each piece may be highly erroneous. These errors can be catastrophic as processing reaches higher levels. A more conservative approach assigns *all* plausible descriptions or labels to each piece and allows a higher-level process to disambiguate the labelings of each piece.

We therefore adopt the position that the shape to be analyzed is decomposed into many, possibly overlapping, primitives, and that each primitive is labeled with *all* plausible terminal symbol names. A *level-0 hypothesis* denotes the association of a single primitive with a single terminal symbol.

The goal of the syntax analysis procedure is to discover all subsets of the set of level-0 hypotheses which can be used to form a parse of the shape. Many of the level-0 hypotheses will not be part of any such grammatical subset; the sooner it is discovered that a level-0 hypothesis (or, as we shall see, a level-$n$ hypothesis) cannot be part of a grammatical description of the shape and is therefore discarded from further consideration, the more efficient in time and space the grammatical analysis will be. Such superfluous hypotheses can be discovered by constraint propagation techniques which apply local constraints on the appearance of a shape to a hierarchical network of hypotheses about the shape. We describe how such local contextual constraints can be generated from a particular class of shape grammars, called *stratified context-free shape grammars.* These constraints are not only derived for terminal symbols, but for symbols at all levels of description.

Thus, the grammar plays two roles:

1. A source of constraints on the local appearance of the shape
2. A model for imposing organization on larger and larger pieces of the shape

The remainder of this chapter is devoted to the description of a shape analysis process, called a *hierarchical constraint process* for integrating these two roles of grammars into one bottom-up, constraint-based parsing procedure. They are an extension of the hierarchical relaxation process described in Davis and Rosenfeld[9]. Here we explain how the desired integration is accomplished, and we only deal with grammar design as a secondary issue, except to the extent that the grammars be of a form from which constraints can be derived.

The hierarchical constraint process can be compared with other proposed syntactic shape analyzers. Pavlidis and Ali[62] suggest that many shape analysis problems can be solved using relatively simple grammars (i.e., finite state) whose terminal symbols are themselves syntactic objects (corners, arcs, etc.) and whose input is a piecewise linear approximation to the shape being analyzed. They describe applications to character recognition and printed circuit board analysis. They do not deal explicitly with the problem of superfluous hypotheses, although their system could be extended to handle this problem.

Tang and Huang[84] do explicitly consider the problem of superfluous hypotheses and propose a solution to the problem based on the design of what they call a "creation machine." A creation machine is an abstract mechanism which applies formal language theory to the construction of strings which are prefixes of the language for the grammar at hand. The syntactic-semantic grammars they describe are similar to the stratified grammars used here. In contrast to the creation machine, the hierarchical constraint process finds locally *ungrammatical* fragments at *all* positions in the shape (rather than just grammatical fragments at the beginning) and propagates the effects of deleting those fragments from the current set of hypotheses about the shape.

Using the hierarchical constraint process involves several steps. A stratified shape grammar must be defined by the user. Certain predicates and functions are available for use in specifying the semantic consistency part and the semantic generator part of a production. Once a shape grammar has been specified, the implicit syntactic and semantic constraints can be derived from the grammar. Finally, the hierarchical constraint process is applied to the data which consists of primitives, i.e., segments of the boundary of a shape. The hierarchical constraint process takes as input the compiled version of the grammar and the primitives and produces as output a network of hypotheses. The network may be empty or may yield several airplane shapes which are found in the data. These ideas are illustrated by a grammar for a simple house shape.

## 7.1 The Grammatical Shape Model

Grammatical models for shape analysis have been developed and investigated by many workers[14,19,20,41,57,61,68,73,81,86]. With some simple modifications, these models can be integrated in a natural way with constraint analysis techniques. An extension of the geometrical grammars of Vamos and Vassy[86] and Gallo[20] will be used to model shapes.

**Definition 7.1** *A* stratified context-free grammar, *G, is a quadruple ( T, N, P, S), where*

- *T is the set of terminal symbols*
- *N is the set of nonterminal symbols*
- *P is the set of productions*
- *S is the set of start symbols.*

Let V = (N $\bigcup$ T) be the set of vocabulary symbols.

Associated with every symbol v $\in$ V is a level number, ln(v) : V $\rightarrow \{0, 1, \ldots, n\}$. $\forall$ v $\in$ T, ln(v) = 0.

1. T corresponds to relatively large pieces of the shapes modeled by the grammar, e.g., straight-edge approximations to the boundary of the shape.

2. N consists of a set of symbols each of which has a level number from 1 to $n$ associated with it. A start symbol has level number $n$, and in any rule v := a (the rewrite part of a production), if ln(v) = $k$, $1 \leq k \leq n$, then every symbol in the string a is at level $k - 1$. Furthermore, for every v $\in$ V

v = <name part> {attachment part} [semantic part]

where:

(a) <name part> is a unique name by which the symbol v is known.

(b) {attachment part} is a set of attachment points of the symbol.

(c) [semantic part] is a set of predicates which describes certain aspects of the symbol.

3. P consists of productions of the form (v:=a, A, C, $G_a$, $G_s$), where

   (a) v:=a is the rewrite part that indicates the replacement of the symbol v by the group of symbols a, where $v \in N$ and $a = v_1 v_2 \ldots v_k$ [$v_i \in V$ and $\ln(v_i) = (\ln(v)-1)$, $i = 1, k$].

   (b) A — set of applicability conditions on the syntactic arrangment of the $v_i, i = 1, k$.

   (c) C — semantic consistency of the $v_i, i = 1, k$, and consists of various predicates describing geometric and other properties of the $v_i$.

   (d) $G_a$ — rules for generating the attachment part for v.

   (e) $G_s$ — rules for generating the semantic part of v.

As an example of the productions of the grammar, consider how engines are formed (see Figure 7.1):

$$\begin{aligned} \langle \text{ engine } \rangle \{ e_1, e_2 \} [ a, \text{span} ] := \; & \langle \text{ engine side } \rangle \{ e'_1, e'_2 \} [ a' ] + \\ & \langle \text{ engine front } \rangle \{ e''_1, e''_2 \} [ a'' ] + \\ & \langle \text{ engine side } \rangle \{ e'''_1, e'''_2 \} [ a''' ] \end{aligned}$$

A : [ Join ($e'_1$ or $e'_2$, $e''_1$) and Join ($e'''_1$ or $e'''_2$, $e''_2$) or Join ($e'''_1$ or $e'''_2$, $e''_1$) and Join ($e'_1$ or $e'_2$, $e''_2$) ]

C : [ Parallel ($a'$, $a'''$) and Length ($a'$) = Length($a'''$) and Perpendicular ($a'$, $a''$) and Parallel ($a''$, Vector(Midpt($a'$), Midpt($a'''$)))]

$G_a$ : [ Set ($e_1$, Unjoined($e'_1$, $e'_2$)) and Set ($e_2$, Unjoined($e'''_1$, $e'''_2$)) or Set ($e_1$, Unjoined($e'''_1$, $e'''_2$)) and Set ($e_2$, Unjoined($e'_1$, $e'_2$)) ]

$G_s$ : [ $a := (a' + a''')/2$ and span := $a''$ ]

This rule specifies that an "engine" is composed of two "engine side" symbols and an "engine front" symbol. A, C, $G_a$, and $G_s$ can be viewed as a program for producing "engine" from symbols on the right-hand side of the rewrite rule. A specifies the physical connections of the symbols on the right-hand side of the rewrite rule, i.e., that each end of the "engine front" has an "engine side" attached to it, but the "engine side" symbols are not connected to each other (see Figure 7.1), where Join (x, y) means that x and y are the same point. C indicates that the two "engine side" symbols attach to it, but the "engine side" symbols should be parallel, of the same length, perpendicular to the "engine front" symbol, and on the same side of the "engine front." $G_a$ and $G_s$ describe the derivation of the attachment points and semantic features for "engine"; the unjoined endpoints of the "engine side" symbols can be given either attachment point name due to the symmetry of the symbol, where Unjoined (x,y) indicates the endpoint which did not satisfy the Join condition in the applicability condition of the production, and Set (x,y) means that

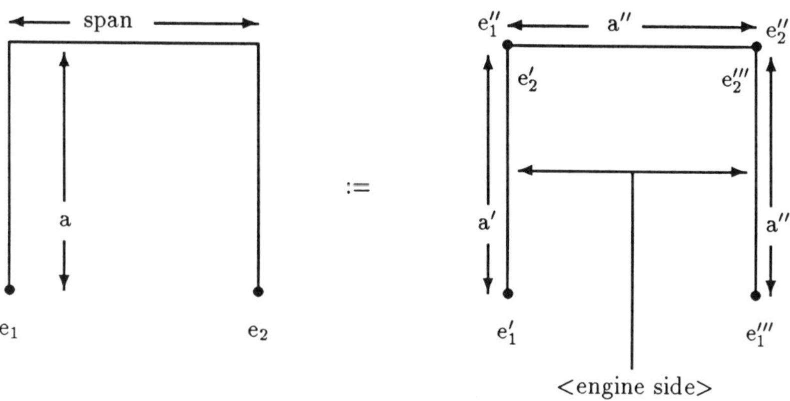


Figure 7.1: Example of a production.

the point x of the hypothesis being created is given the physical attributes of the existing endpoint y. The main axis is the average of those of the "engine side" symbols, and the span is exactly that of "engine front."

As a simple example for a complete grammar, consider G = (T,N,S,P), where

T = { roof, wall, floor }
N = { top, bottom, house }
S = { house }
P = { $\text{production}_1$, $\text{production}_2$, $\text{production}_3$ }

$\text{production}_1$:

<top> $\{e_1,e_2\}$ [a] := <roof> $\{e_1', e_2'\}$ [$a'$] +
<roof> $\{e_1'', e_2''\}$ [$a''$]

A : [ Join ($e_1', e_1''$) or Join ($e_1', e_2''$) or Join ($e_2', e_1''$) or Join ($e_2', e_2''$) ]

C : [ Perpendicular ($a', a''$) and Equal ( Length ($a'$), Length ($a''$) ) ]

$G_a$ : [ Set ($e_1$, Unjoined ($e_1', e_2'$)) and Set ($e_2$, Unjoined ($e_1'', e_2''$) or
Set ($e_1$, Unjoined ($e_1''$), $e_2''$) and Set ($e_2$, Unjoined ($e_1', e_2'$)) ]

$G_s$ : [ Set (a, Vector ($e_1, e_2$)) ]

$\text{production}_2$:

<bottom> $\{e_1, e_2\}$ [a] := <wall> $\{e_1', e_2'\}$ [$a'$] +
<floor> $\{e_1'', e_2''\}$ [$a''$] +
<wall> $\{e_1''', e_2'''\}$ [$a'''$]

A : [ Joined ($e_1', e_1''$) and Joined ($e_2'', e_1'''$) or Joined ($e_1', e_1''$) and
Joined ($e_2'', e_2'''$) or Joined ($e_1', e_2''$) and Joined ($e_1'', e_1'''$) or

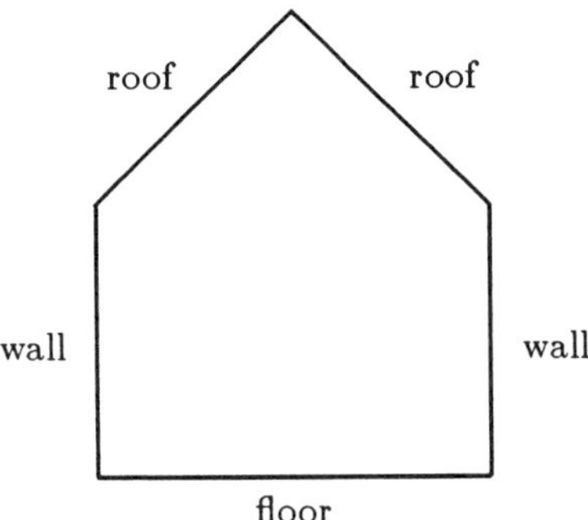


Figure 7.2: Typical house shape.

Joined ($e'_1$, $e''_2$) and Joined ($e''_1$, $e'''_2$) or Joined ($e'_2$, $e''_1$) and Joined ($e''_2$, $e'''_1$) or Joined ($e'_2$, $e''_1$) and Joined ($e''_2$, $e'''_2$) or Joined ($e'_2$, $e''_2$) and Joined ($e''_1$, $e'''_1$) or Joined ($e'_2$, $e''_2$) and Joined ($e''_1$, $e'''_2$) ]

C : [ Parallel (a', a''') and Perpendicular (a', a'') and Equal ( Length (a'), Length (a''), Length (a''') ) ]

$G_a$ : [ Set ($e_1$, Unjoined ($e'_1$, $e'_2$)) and Set ($e_2$, Unjoined ($e'''_1$, $e'''_2$)) or Set ($e_1$, Unjoined ($e'''_1$, $e'''_2$)) and Set ($e_2$, Unjoined ($e'_1$, $e'_2$)) ]

$G_s$ : [ Set (a, a'') ]

production$_3$:

<house> { } [a] := <top> {$e'_1$, $e'_2$} [a'] + <bottom> {$e''_1$, $e''_2$} [a'']

A : [ Joined ($e'_1$, $e''_1$) and Joined ($e'_2$, $e''_2$) or Joined ($e'_1$, $e''_2$) and Joined ($e'_2$, $e''_1$) ]

C : [ Equal (a', a'') and Parallel (a', a'') ]

$G_a$ : [ ]

$G_s$ : [ Set (a, a') ] }

Figure 7.2 shows a typical example of the shapes described by G.

Stratified grammars naturally give rise to a large set of contextual constraints on the organization of shape. It is these constraints which the hierarchical constraint process utilizes to analyze shapes.

## 7.2 Compilation of Constraints

We now discuss the procedures for deriving the local constraints from the shape grammar. Two types of constraints, syntactic and semantic, are described. The semantic attributes of a vocabulary symbol are computed from the attributes of the symbols which produce it (see Knuth[45] for a discussion of defining semantics for context-free languages using both synthesized and inherited attributes; we use only synthesized attributes here). Consider a vocabulary symbol as representing a piece of the boundary of a shape. If a vocabulary symbol is part of a complete shape, then it is adjacent to pieces of the shape which can combine with it to produce a higher-level vocabulary symbol. Therefore, if the set of all possible neighbors of a vocabulary symbol is known, and at one of its attachment points no hypothesis for any of these symbols exists, then that hypothesis can be eliminated. This type of constraint is called a syntactic constraint. Without these constraints several levels of vocabulary symbols might be built before it can be determined that some hypothesis lacks the appropriate context. The use of constraints, however, makes it possible to detect much earlier the lack of appropriate context. The other type of constraint involves some geometric relation between the semantic features of two vocabulary symbols, e.g., the main axis of a <plane> is parallel to the main axis of an <engine>. These kinds of constraints are called semantic constraints. This makes it possible for high-level information to be specified, e.g., the orientation of the plane, and for this information to be used to delete hypotheses which are not consistent with the given information, e.g., the main axis of any engine hypothesis must be parallel to the orientation of the plane.

## 7.3 Compiling Syntactic Constraints

Let G = (T,N,P,S), let v, w and x $\in$ V, let at(v) denote the attachment points of v, and let $a_v \in$ at(v). We define,

**Definition 7.2** $v$ ancestor$_{a_v,a_w}$ $w$ *iff* $\exists\ p \in P \ni$ *the rewrite rule of* $p$ *is* $v := \ldots \mathrm{w} \ldots$ *and* $\exists\ a_w \in at(w) \ni a_w$ *is identified with* $a_v \in G_a$ *of* $p$. *That is, the attachment point* $a_v$ *of the left-hand side symbol,* $v$*, is associated with endpoint* $a_w$ *of the right-hand side symbol,* $w$*. For example, in Figure 7.1 the attachment points for the symbol "engine" are associated with the unjoined attachment points of the "engine side" symbols.*

**Definition 7.3** $w$ descendent$_{a_w,a_v}$ $v$ *iff* $v$ *ancestor*$_{a_v,a_w}$ $w$.

**Definition 7.4** $v$ neighbor$_{a_v,a_w}$ $w$ *iff*

1. $\exists\ p \in P \ni$ *the rewrite rule of* $p$ *is* $x := \ldots \mathrm{v} \ldots \mathrm{w} \ldots$ *and* $a_w$ *is specified as being joined to* $a_v$ *in the applicability conditions, A, of* $p$*, or*

2. $\exists\ x \in V$ *with* $a_x \in at(x)$*, and* $\exists\ y \in V$ *with* $a_y \in at(y) \ni x$ *ancestor*$_{a_x,a_v}$ $v$*, and* $y$ *neighbor*$_{a_y,a_x}$ $x$*, and* $w$ *descendent*$_{a_w,a_y}$ $y$. *Note that computing the neighbor relation for level k symbols assumes knowing the neighbor relation for all levels greater than k.*

Using matrix representations for these relations, the descendents and neighbors of a symbol at a particular attachment point can be computed (see Gries[25] for an introduction to binary relations, their representation using matrixes, and their manipulation). The notation w $R_{a_w,a_v}$ v indicates that w is in relation R to v through endpoint $a_w$ of w and $a_v$ of v. Given $k$ attachment points per vocabulary symbol, the $\text{neighbor}_{i,j}$ relation (assuming an ordered set of attachment points and i, j are in the range 1 to $k$) is computed by iterating the following computation $n-1$ times:

$$\text{neighbor}_{i,j} := \text{neighbor}_{i,j} + \sum_{m=1}^{k} \left\{ \text{descendent}_{i,m} \times \left[ \sum_{p=1}^{k} \left( \text{neighbor}_{m,p} \times \text{ancestor}_{p,j} \right) \right] \right\}$$

Consider the house grammar given in Section 7.1. An ancestor matrix $\mathbf{M}_{ij}$ is a square matrix whose order is the number of vocabulary symbols in the grammar, and for which $i$ specifies the attachment point of the vocabulary symbol of the row, and $j$ specifies the attachment point of the vocabulary symbol of the column. Since there are two attachment points for each symbol of the grammar, then there should be four matrixes specifying the ancestor relation; however, as the attachment conditions and the attachment part generator are symmetric, all four ancestor matrices are equal:

$$\mathbf{A} = \mathbf{A}_{11} = \mathbf{A}_{12} = \mathbf{A}_{21} = \mathbf{A}_{22} = \begin{bmatrix} 0 & 0 & 0 & 0 & 0 & 0 \\ 0 & 0 & 0 & 0 & 0 & 0 \\ 0 & 0 & 0 & 0 & 0 & 0 \\ 1 & 0 & 0 & 0 & 0 & 0 \\ 0 & 1 & 0 & 0 & 0 & 0 \\ 0 & 0 & 0 & 0 & 0 & 0 \end{bmatrix}$$

where the rows and columns 1 through 6 correspond to the vocabulary symbols <roof>, <wall>, <floor>, <top>, <bottom> and <house>, respectively. Note that the <house> symbol could be left out as it has no attachment points. The descendent matrix $\mathbf{D}_{ij}$ is just the transpose of $\mathbf{A}_{ij}$:

$$\mathbf{D} = \mathbf{D}_{11} = \mathbf{D}_{12} = \mathbf{D}_{21} = \mathbf{D}_{22} = \begin{bmatrix} 0 & 0 & 0 & 1 & 0 & 0 \\ 0 & 0 & 0 & 0 & 1 & 0 \\ 0 & 0 & 0 & 0 & 0 & 0 \\ 0 & 0 & 0 & 0 & 0 & 0 \\ 0 & 0 & 0 & 0 & 0 & 0 \\ 0 & 0 & 0 & 0 & 0 & 0 \end{bmatrix}$$

Finally, the explicit neighbor relation is given as:

$$\mathbf{N} = \mathbf{N}_{11} = \mathbf{N}_{12} = \mathbf{N}_{21} = \mathbf{N}_{22} = \begin{bmatrix} 1 & 0 & 0 & 0 & 0 & 0 \\ 0 & 0 & 1 & 0 & 0 & 0 \\ 0 & 1 & 0 & 0 & 0 & 0 \\ 0 & 0 & 0 & 0 & 1 & 0 \\ 0 & 0 & 0 & 1 & 0 & 0 \\ 0 & 0 & 0 & 0 & 0 & 0 \end{bmatrix}$$

which is determined by the grammar. The full neighbor relation is computed as:

$$\mathbf{N} := \mathbf{N} = \mathbf{DNA} = \begin{bmatrix} 1 & 1 & 0 & 0 & 0 & 0 \\ 1 & 0 & 1 & 0 & 0 & 0 \\ 0 & 1 & 0 & 0 & 0 & 0 \\ 0 & 0 & 0 & 0 & 1 & 0 \\ 0 & 0 & 0 & 1 & 0 & 0 \\ 0 & 0 & 0 & 0 & 0 & 0 \end{bmatrix}$$

The full neighbor relation includes the relation between <roof> and <wall> which was absent in the neighbors taken straight from the grammar. Each row gives the set of vocabulary symbols which can possibly neighbor the symbol associated with that row. If a hypothesis fails to have a neighbor from this set, then that hypothesis can be deleted.

## 7.4 Compiling Semantic Constraints

Semantic constraints can be generated in exactly the same way as syntactic constraints, i.e., by defining binary relations and compiling their transitive closure. For example, the axes of two symbols are parallel if a production states this explicitly, or if each symbol has an ancestor parallel to itself and these ancestors are explicitly parallel. There are two methods for dealing with semantic constraints.

First, associate with each endpoint of a vocabulary symbol an end angle. Then, in either the semantic consistency or the applicability condition indicate an allowed range of angle for two symbols to have if joined at particular endpoints.

This approach is analogous to the syntactic neighbor case; now a relation is defined between every two symbols joined at particular endpoints and the closure will give results for implicit symbol neighbors (i.e., even if two symbols do not appear together in the right-hand side of some production, but are joined at a higher level). There are several problems with doing things this way.

In practice the range of the angle is too large to give much extra constraint, and the explicit semantic constraints in the productions are adequate. A more serious problem concerns making use of the semantic constraints in a more global way. Since a range of angle is given at every endpoint, then if the allowed range is computed for symbols several neighbors away, the allowed range is useless as a constraint.

Alternatively, we can design special-purpose relations which when computed will yield semantic constraints between vocabulary symbols. Such constraints can be more global in nature and can be applied with much more certainty. For example, the parallel relation can be used to delete an <engine side> hypothesis that is not parallel to some other <engine side> hypothesis. Such relations also allow for semantic features to be fixed (set to some constant, e.g., the orientation of the main axis of a plane could be set to 45 degrees), and this certain information can be propagated through the network of hypotheses. This can be done, for example, by having global information available describing known orientations of the vocabulary symbols. In this way, it will be possible to determine which hypotheses can be deleted.

Nothing precludes the use of both methods; however, due to the reasons given above, only the special-purpose relations were used. In particular, the parallel relation was computed between all vocabulary symbols. Note that not every symbol is necessarily parallel to another symbol, and as pointed out above, some hypothesized symbols may require other hypotheses of the same symbol to exist, e.g., <engine side>.

The parallel relation can occur between the axes of two vocabulary symbols in a variety of ways:

1. They can be explicitly defined as parallel in the semantic consistency part of a production.

2. The semantic generation part of a production may set an axis of the new vocabulary symbol equal to an axis of one of the vocabulary symbols being used to produce the new symbol.

3. They may be indirectly parallel if there exists a third vocabulary symbol to which they are both parallel.

For example, the <distal wing edge> is parallel to the <engine front> through the <distal wing edge1> (i.e., across level symbols are parallel, and this can be used to find symbols at a lower level which are parallel). These relations are computed using a binary-valued matrix, whose rows and columns correspond to the axes of the vocabulary symbols.

Again consider the grammar for the simple house shape. Let the matrix **P** denote the parallel relation between the axes of the vocabulary symbols. Since only one axis exists per vocabulary symbol, only one matrix is necessary to define the parallel relation. The matrix given directly by the grammar is:

$$\mathbf{P}' = \begin{bmatrix} 0 & 0 & 0 & 0 & 0 & 0 \\ 0 & 1 & 0 & 0 & 0 & 0 \\ 0 & 0 & 0 & 0 & 1 & 0 \\ 0 & 0 & 0 & 0 & 1 & 1 \\ 0 & 0 & 1 & 1 & 0 & 0 \\ 0 & 0 & 0 & 1 & 0 & 0 \end{bmatrix}$$

The transitive closure of $\mathbf{P}'$ is:

$$\mathbf{P} = \begin{bmatrix} 0 & 0 & 0 & 0 & 0 & 0 \\ 0 & 1 & 0 & 0 & 0 & 0 \\ 0 & 0 & 0 & 1 & 1 & 1 \\ 0 & 0 & 1 & 0 & 1 & 1 \\ 0 & 0 & 1 & 1 & 0 & 1 \\ 0 & 0 & 1 & 1 & 1 & 0 \end{bmatrix}$$

except that elements on the diagonal of $\mathbf{P}'$ remain unaltered, where $\mathbf{P} = (\mathbf{P}' + \neg\mathbf{I}) * \mathbf{P}'!$, where $\mathbf{I}$ is the Boolean identity matrix and $\mathbf{P}'!$ is the transitive closure of $\mathbf{P}'$. This must be done since by transitivity a vocabulary symbol would be parallel to itself through any other element to which it is parallel. For example, any <wall> hypothesis must have a distinct <wall> hypothesis parallel to it. However, this constraint was already explicit in the productions; a less obvious constraint is that any <top> symbol must be parallel to a <floor> symbol. Since <floor> is a lower-level symbol, all <floor> symbols will already have been built by the time <top> symbols are being built, and this can be used to delete <top> hypotheses which are not parallel to any <floor> hypothesis. For example, suppose the level 0 hypotheses are as given in Figure 7.3. All the syntactic constraints are satisfied. Hypotheses (1,2) and (2,3) satisfy the conditions for making the level 1 hypothesis ¡top¿, but since there is no level 0 hypothesis which is parallel to the ¡top¿ axis, the ¡top¿ hypothesis will be deleted.

In order to add other semantic constraints, it is necessary to have a matrix to define the constraints. The matrix can be computed from the grammar once the relations have been defined in terms of the predicates which appear in the productions. Parallel is a transitive relation, and other transitive relations can be computed in much the same way as was done for parallel. Relations which are not transitive, e.g., perpendicular, require special procedures for their computation.

Some applications may prohibit the precompilation of all constraints, e.g., due to the size of the grammar. In such cases a possible alternative is to compute relations only when necessary. For example, the investigation of the relations between low-level hypotheses could be guided by the incorporation of some top-down structure for the application of HCP to shape analysis. Such an approach would restrict the number of symbols for which constraints would be required; namely, only those constraints concerning symbols explored by the top-down parse would be computed. Of course, once the relations between the features of two vocabulary symbols have been computed, they can be stored for future reference and need not be recomputed every time.

## 7.5 The Hierarchical Constraint Process

As discussed in the introduction, a major problem associated with syntactic pattern recognition is the segmentation of the object into pieces which correspond to the terminal symbols of the grammar. A high false alarm rate implies that many primitives will be generated, and correspondingly many terminal symbols hypothesized from them, this implying a large search space. In order to overcome these difficulties, a *hierarchical constraint process* (HCP) uses hierarchical models of objects and uses model-derived constraints to eliminate inconsistent hypotheses at each level of the model. In particular, using the stratified context-free grammars already described, syntactic (e.g., spatial concatenation) and semantic (e.g., symmetry, collinearity, etc.) constraints can be automatically generated to guide the analysis of the shape.

Primitives for the grammatical analysis are generated by computing several piecewise linear approximations to the boundary of the shape. A modified split-and-merge algorithm fits straight edges to the boundary using the corner measure proposed by Freeman and Davis[15] to choose break points. For each point on the original boundary, an error measure defined as the minimum distance from that boundary segment containing that boundary point is computed. Then, an error measure for the line segment is defined to be the sum of the errors of each underlying boundary point. Primitives are generated at various error thresholds; even though stricter thresholds are applied to segmentations already generated, this is not the same as applying the stricter thresholds to each segment which would result in a hierarchical set of primitives. Neighboring primitives from different thresholds are not guaranteed to have endpoints that meet exactly, but primitives are now defined as neighbors if they have endpoints that lie within some disk of fixed size. By computing several segmentations, it is hoped that all the necessary primitives will be found. The search will be made feasible by the constraints implied in the grammar and imposed by the relaxation techniques.

The association of terminal symbols with primitives will (in the limit) be to hypothesize every terminal for each primitive. However, methods for reducing the number of hypotheses include using a more global analysis to derive indications of appropriate scale, orientation, etc. from simple global properties, e.g., histogramming selected features of the primitives themselves and using the model to infer properties of particular terminal symbols.

The hierarchical relaxation system computes a bottom-up parse of the shape by applying the constraints to a network of low-level hypotheses about pieces of the shape. The processing of this network can be easily described by specifying three simple procedures and two sets which these procedures manipulate:

- BUILD: Given level $k$ of the network, BUILD uses the productions of the grammar to construct nodes corresponding to level $k+1$ node as supporting it, and it, in turn, is recorded as supported by them. After all nodes are generated, node pairs corresponding to adjacent boundary segments are linked only if the constraints allow the symbols hypothesized for that pair to be adjacent. Any level $k$ symbols which are used to generate a node at level $k+1$ are associated with that level $k+1$ symbol. Building level 0 involves applying the segmentation strategy to the shape to generate the level 0 nodes.

- CONSTRAIN: Since each node corresponds to a single hypothesis, and since nodes are only linked to compatible nodes, the within layer application of syntactic constraints simply involves removing a node if it has no neighbor at some endpoint. Likewise, a node is removed if any semantic constraints are not satisfied.

- COMPACT: Given a node $n$ at level $k$, if level $k-1$ has been built and $n$ does not support a level $k+1$ node, then $n$ is deleted from the network. If any of the nodes which produced n have been deleted, then $n$ is deleted, too.

These procedures operate on two sets of nodes, $R_x$ and $R_c$, both of which are initially empty. When at level $k$ with $R_x$ and $R_c$ empty, BUILD produces the level $k+1$ hypotheses (or stops if $k = n$), and puts them into $R_x$ while putting all level k nodes into $R_c$. CONSTRAIN then removes nodes from $R_x$, taking no action if the node has a neighbor at all endpoints, but otherwise deleting the node from the network and putting its same level neighbors in $R_x$ and its across level neighbors in $R_c$. COMPACT removes nodes from $R_c$, taking no action if all the node's original supporting nodes still exist at level $k-1$ and the node still supports at least one level $k+1$ node (if level $k+1$ as been built); otherwise, COMPACT deletes the node from the network and puts its same level neighbors in $R_x$ and its across level neighbors in $R_c$.

HCP does not eliminate any hypothesis which contributes to a complete parse. This can be seen as follows. BUILD simply generates the next level symbols, and if used without CONSTRAIN and COMPACT, will produce all possible hypotheses at every level. CONSTRAIN is applied to a set of nodes taken one at a time, and if a hypothesis h is deleted, it is precisely for the reason that at one endpoint of h no neighboring hypotheses can be joined to h to produce a higher-level symbol. As for COMPACT, there are two cases to consider. First, if a level $k$ hypothesis is not used to produce any level $k+1$ hypothesis, then since level $k+1$ is only built once, that level $k$ hypothesis will never produce any higher-level hypothesis. Thus, eliminating it does not affect any complete parse. Finally, if a level $k$ hypothesis h loses the support of one or more of the hypotheses which produced it, then clearly if h were part of some complete parse, then the supporting nodes would be, too.

Consider once more the house grammar. Suppose level 0 hypotheses for the seven primitives are as given in Figure 7.3, where r, w, f, t, b, and h stand for <roof>, <wall>, <floor>, <top>, <bottom> and <house>, respectively, and every possible hypothesis has been made for each primitive. The syntactic and semantic constraints are as given in Sections 7.2 and 7.3. Figure 7.4 shows the network of hypotheses as an *adjacency graph*, that is, the nodes correspond to level 0 hypotheses, and given two nodes, $n_1$ and $n_2$, $n_1$ is connected by a directed arc to n if the primitive corresponding to $n_1$ in the shape "precedes" and is adjacent to the primitive corresponding to $n_2$ (precedes is well-defined if we adopt a specific sense

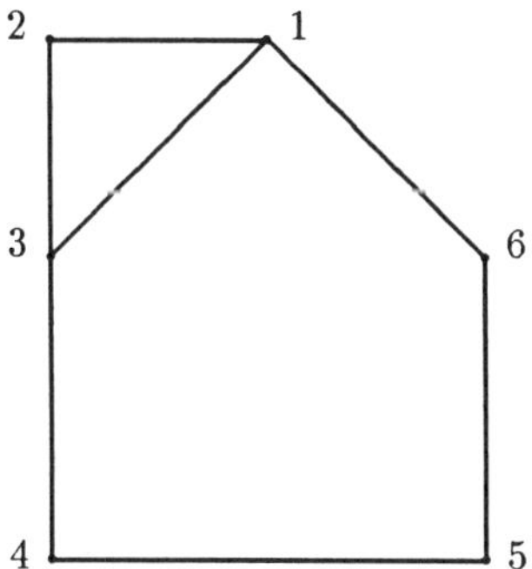


Figure 7.3: House primitives (1,2), (1,3), etc. and their hypotheses.

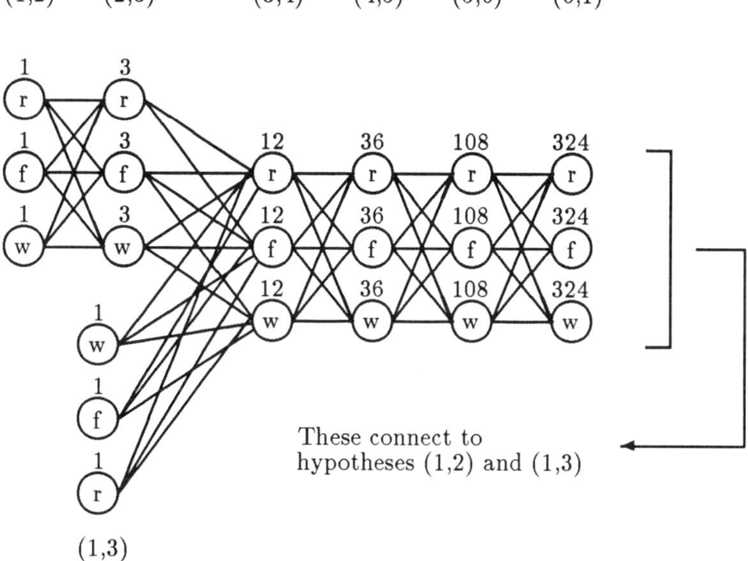


Figure 7.4: Corresponding network of hypotheses.

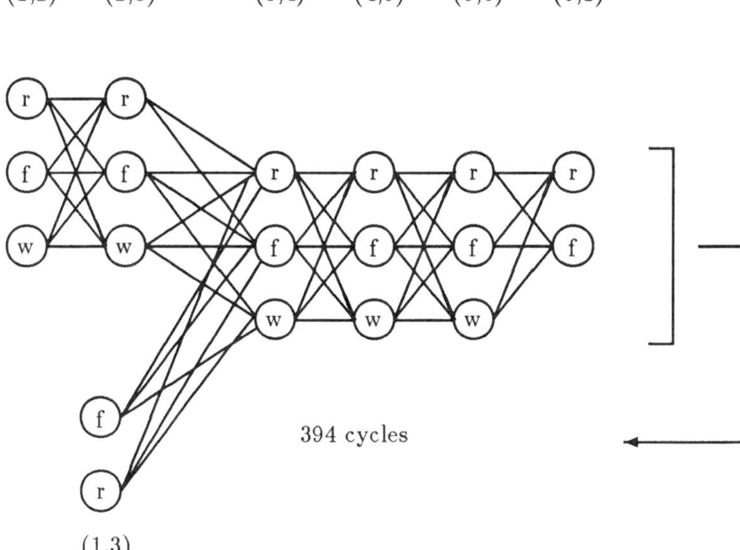


Figure 7.5: Hypotheses after semantic constraints.

for following the shape boundary). The number next to each node is the number of distinct paths (from left to right) to that node from a leftmost node. This number is computed for nodes from left to right by assigning the leftmost nodes a value of 1, and continuing to the right assigning to each node the sum of those nodes immediately to the left and adjacent to it. Each rightmost node connects to each leftmost node for a closed shape. Note that the leftmost nodes may derive from distinct primitives that all share a common endpoint. The goal of HCP is to find all the cycles in the network that correspond to shapes in the language defined by the grammar.

HCP begins by applying CONSTRAIN to the network of hypotheses: All the syntactic constraints are satisfied; however, the <wall> hypotheses of primitives (1,3) and (6,1) are deleted since neither of them has a parallel <wall> hypothesis. The result is shown in Figure 7.5. The syntactic constraints now cause several hypotheses to be deleted: The <floor> hypotheses are deleted from (1,2), (1,3), and (5,6) since one attachment point fails to have a neighboring <wall> hypothesis. Figure 7.6 shows the result.

At this point, all syntactic and semantic constraints are satisfied, and all possible level 1 vocabulary symbols can be built. Each hypothesis is checked against all applicable productions in building the next level. For example, the primitives in Figure 7.6 can be combined in the following way: The (1,2) <roof> hypothesis can be joined with the (2,3) <roof> hypothesis to form a <top> hypothesis; however, the axis of the resulting <top> hypothesis has no <floor> hypothesis parallel to it, and this <top> hypothesis is therefore discarded. The (1,2) <roof> hypothesis

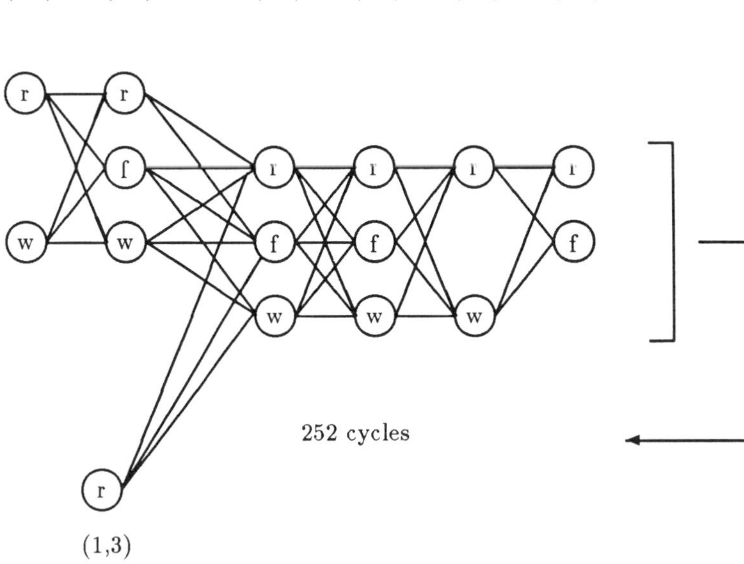


Figure 7.6: Hypotheses after syntactic constraints.

and the (1,6) <roof> hypothesis, although connected properly, fail the semantic consistency part of the production as they are not perpendicular. The other hypotheses are investigated in a similar manner, and the resulting level 1 hypotheses are shown in Figure 7.7. COMPACT is now applied to the level 0 hypotheses, and the result is shown in Figure 7.8. At this point, all incorrect hypotheses have been eliminated, and the level 2 start symbol <house> is easily built. Even this simple example shows the advantage of using syntactic and semantic constraints.

Of course, constraints can be generated from grammars that are not stratified, but the application of the constraints will not prevent the repeated production of symbols which fail to satisfy the constraints. This is due to the fact that a hypothesis cannot be discarded since it could be used at any time. However, stratification insures that a symbol will be built only once.

## 7.6 Shape Analysis Using HCP

The input to HCP consists of the compiled grammar, i.e., the productions of the grammar and the derived constraints, and a set of primitives. Section 7.4 describes how the constraints are determined. The boundary segments are produced by a split-and-merge algorithm. The split-and-merge algorithm is applied at several thresholds so as to produce various levels of "smoothing" of the original data.

Each primitive has several hypotheses associated with it due to the ambiguity of line segments at the lowest level of shape description, i.e., most detailed pieces of the shape can be of any length or orientation. In order to be certain that a shape is found, it is necessary to hypothesize every level 0 vocabulary symbol for every primitive. Since this may be quite a large set, several procedures have been

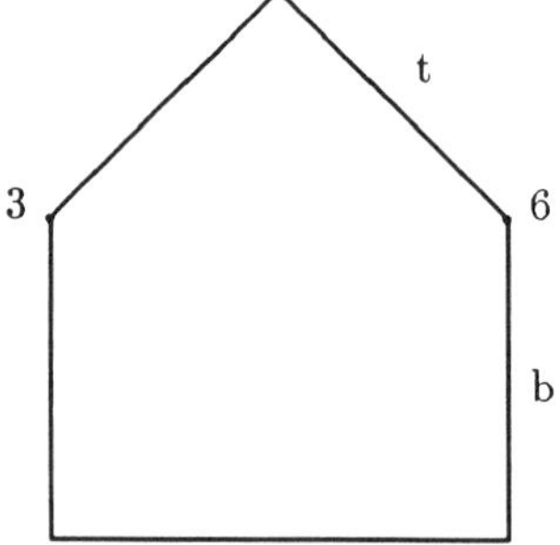


Figure 7.7: Level 1 of house.

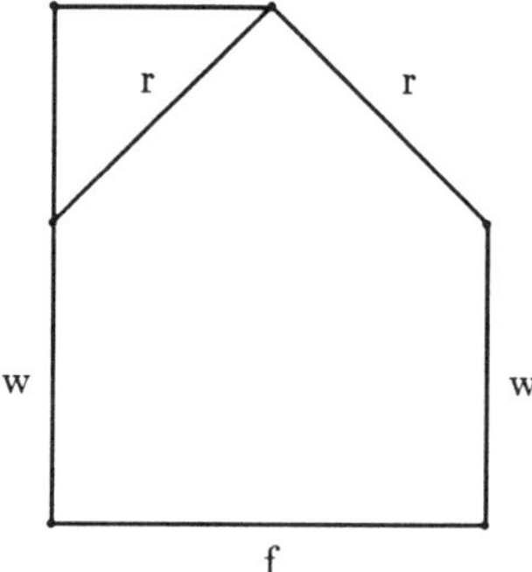


Figure 7.8: Final level 0 hypotheses.

developed for reducing its size. One approach is to use the statistics of the geometric properties of the primitives taken as a group. For example, the histogram of the lengths of the primitives can be used to rule out certain hypotheses for a particular primitive, e.g., <wing tip> is most probably not a good hypothesis for the longest primitives. In any case, once an initial set of hypotheses for each primitive has been determined, a network of hypotheses can be constructed.

Each hypothesis in the network consists of several parts called *node parts*:

- NAME
- ATTACHMENT POINTS
- SEMANTIC FEATURES
- PRIMITIVE
- LEVEL NUMBER
- ASSOCIATED VOCABULARY SYMBOL
- NEIGHBORS
- SUPPORTS
- SUPPORTED BY
- LINK

Each node has a unique name and corresponds to a hypothesis about a particular piece of the boundary of the shape. The attachment points of a node specify particular points at which a node can be joined to other nodes to produce a higher-level node. The semantic features include special axes or features of a node, e.g., a <wing> has an axis for both width and breadth and an engine count. Features may have subfeatures themselves, as for example an axis has both a direction and a length. The primitive part of a node specifies the part of the boundary with which a node is associated by giving the (x,y) coordinates of the endpoints. The level number of a node is simply the level number of the associated vocabulary symbol which is also specified.

The node parts discussed so far have only described the node itself. The rest of the node parts give the relation of the node to other nodes in the network. Neighbors of a node are other nodes at the same level which share an endpoint. Thus, a node may have several neighbors at one endpoint for two reasons:

1. The underlying primitive of the node is adjacent to several distinct primitives at that endpoint (this is possible due to the multiple segmentations).
2. An adjacent primitive has several associated hypotheses.

Finally, a node is supported by the nodes which were used to produce it, and a node supports any node it was used to produce.

Given the initial level 0 hypotheses associated with each primitive, a network is constructed in the following way:

1. A node is created for each hypothesis.
2. When all these nodes have been built, the neighbors of each node are determined (by checking the primitive).

3. The supports and supported by lists are set to nil.

Note that the only semantic features of the level 0 nodes correspond to the axes of the primitives.

Given this network, HCP is run as outlined in Section 7.5. After BUILD is applied to obtain level $k$, the level $k-1$ nodes are put in the COMPACT set, $R_c$, and the level $k$ nodes on the CONSTRAIN set, $R_x$. CONSTRAIN and COMPACT are run until the sets are empty. These steps are repeated until the highest level has been built.

In building level $k$, each node of the network is tried in turn, and every production checked to see if the node's associated vocabulary symbol is in the right-hand side of the production. A vocabulary symbol is composed of several parts:

- NAME PART: The NAME PART is the unique name of the vocabulary symbol.

- FORMAL ATTACHMENT PART: The FORMAL ATTACHMENT PART gives formal names to the attachment points. This is necessary since each vocabulary symbol in the right-hand side must have its own unique names for its attachment points (due to the applicability condition).

- FORMAL SEMANTIC PART: Likewise, the FORMAL SEMANTIC PART allows the use of distinct vocabulary symbols with the same name to be uniquely identified. For example, the production for <engine> has two <engine side> symbols on the right-hand side.

- LEVEL NUMBER: The LEVEL NUMBER is that of the vocabulary symbol.

- SEMANTIC CONSTRAINTS: The SEMANTIC CONSTRAINTS part describes the semantic constraints which exist between a vocabulary symbol and the other vocabulary symbols. This includes a description of the constraint relation (e.g., parallel), the vocabulary symbol feature involved (e.g., a particular axis), the other vocabulary symbols by which it is constrained, and the particular semantic feature of those other vocabulary symbols.

- NEIGHBORS: The NEIGHBORS part of a vocabulary symbol gives the syntactic constraints by endpoint, i.e., a vocabulary symbol can be adjacent to specific symbols at a given endpoint.

- LINK: Pointer to next element in list.

Productions consist of several parts, also:

- NAME: Each production has a unique NAME PART.

- LEVEL NUMBER: The LEVEL NUMBER is that of the vocabulary symbol in the left-hand side.

- LEFT-HAND SIDE: The LEFT-HAND SIDE is the vocabulary symbol produced by the production.

- RIGHT-HAND SIDE: The RIGHT-HAND SIDE is a list of the vocabulary symbols used to produce the symbol in the left-hand side.

- APPLICABILITY CONDITION: The APPLICABILITY CONDITION describes the possible ways that the attachment points of the vocabulary symbols on the right-hand side can be joined together.

- SEMANTIC CONSISTENCY: The SEMANTIC CONSISTENCY gives the necessary relations and predicates that the vocabulary symbols on the right-hand side must fulfill.

- ATTACHMENT PART GENERATOR: If the vocabulary symbols on the right hand side can be joined to produce the vocabulary symbol on the left-hand side, the ATTACHMENT PART GENERATOR describes what the attachment points of the new vocabulary symbol will be.

- SEMANTIC PART GENERATOR: If the vocabulary symbols on the right hand side can be joined to produce the vocabulary symbol on the left-hand side, the SEMANTIC PART GENERATOR describes what the semantic features of the new vocabulary symbol will be.

- LINK: Pointer to the next production.

The APPLICABILITY CONDITION, SEMANTIC CONSISTENCY, ATTACHMENT PART GENERATOR, and SEMANTIC PART GENERATOR are given in terms of predicates:

- NAME: The NAME of a predicate not only uniquely identifies the predicate, but also defines the type of the predicate as "function" or "constant." Only predicates of type function can have a non-nil ARGUMENT part.

- VALUE: The value of the predicate.

- LINK: The LINK part is a pointer to a predicate. Any predicate can have a non-nil LINK part.

- ARGUMENTS: The ARGUMENTS part is a pointer to a predicate.

For example, the expression: $z = (x \cdot x) + (y \cdot y)$ can be expressed as in Figure 7.9. This predicate structure allows for multiple interpretations. That is, depending on the information needed by HCP, the predicate structure can be evaluated as:

1. A proposition having a value of "true" or "false"

2. An action to be taken when generating values for the instantiation of a new hypothesis.

The applicability condition and semantic consistency parts evaluate predicate structures as true or false, while any part of a production can evaluate a predicate structure as a value. The predicate structure given in Figure 7.9 when viewed as a predicate evaluates to "false," and when used to generate a value for $z$, sets the value of $z$ to 500. This allows flexibility both in writing grammars and in using HCP. Two stacks are maintained for evaluating such expressions (the operator stack and the operand stack). Thus, a higher level node is produced if nodes exist that match the right hand side of some production and those nodes satisfy the necessary conditions.

Pseudo-PASCAL versions of HCP, CONSTRAIN, and COMPACT are:

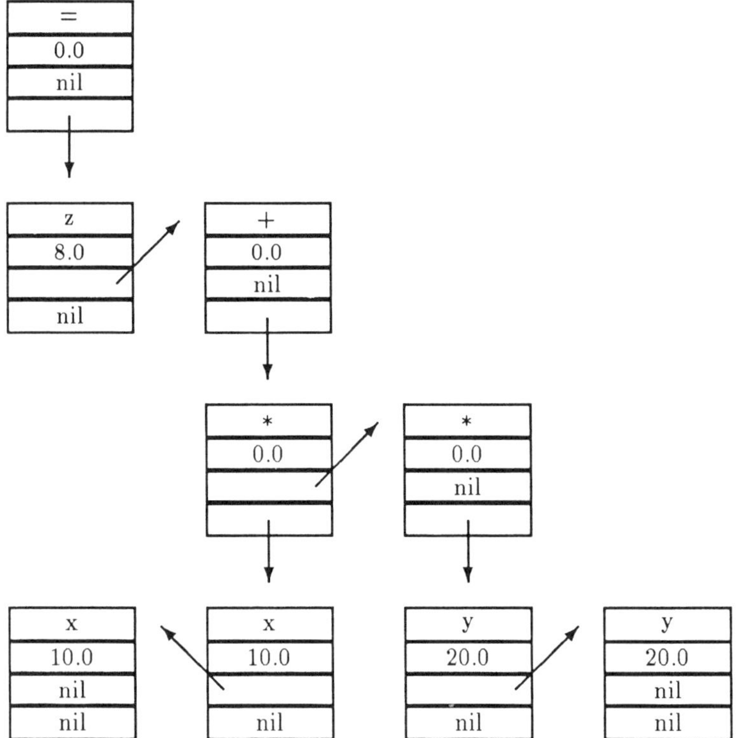


Figure 7.9: Predicate structure.

```
procedure HCP;
begin
 constrain-set := compact-set := ∅;
 level := −1;
 while level < n do begin
  level := level + 1;
  Build( level );
  while (constrain-set ≠ ∅) or (compact-set ≠ ∅) do begin
   Constrain( constrain-set );
   Compact( compact-set );
   end;
  end;
end;

procedure Constrain( constrain-set );
bf while (constrain-set ≠ ∅) do begin
 node := Next-in-set( constrain-set );
 if ¬ Satisfies( Node ) then begin
  Delete-from-network( node );
  Put-on-constrain( Neighbors( node ));
  Put-on-compact( Across-level-neighbors( node ));
  end;
```

```
  end;

procedure Compact ( compact-set );
while compact-set ≠ ∅ begin
  node := Next-in-set( compact-set );
  if ¬ Supportable( node ) then begin
    Delete-from-network( node );
    Put-on-constrain( Neighbors( node ) );
    Put-on-compact( Across-level-neighbors( node ) );
    end;
  end;
```

## 7.7 Strategy Trees

In analyzing shape, more knowledge may be available than can be specified in the hierarchical relaxation process. For example, if the orientation of the main axis of an airplane is known, then this can be incorporated into the semantic constraints and used to eliminate incorrect hypotheses at an early stage of the analysis. Moreover, it is sometimes useful to be able to restrict the grammar to producing particular vocabulary symbols (e.g., wings) instead of always producing the start symbol. That is, it is desirable to have some means of specifying a strategy for using HCP in the given problem context.

This is accomplished by specifying a strategy tree. A strategy tree is a generalization of a hierarchical classifier or decision tree[49]. Decision trees are generally used in conjunction with a statistical pattern recognition scheme. The purpose of using such a device is to avoid computing all possible features. It allows the user to take advantage of known relations between the features. For example, in studying multispectral data, the value of a feature computed on one channel may make it possible to avoid measuring that same feature on the other channels. Decision trees provide a way of ordering the sequence of feature measurements such that at each node of the tree certain classes are no longer considered possible as labels for the test sample. Thus, a decision tree directs overall strategy and permits a final decision to be found rapidly.

We will consider only binary decision trees, that is, a feature will evaluate to true or false and one of two branches will be taken as a result. One weakness of such decision trees is that decisions are irrevocable; robustness can be achieved only by greatly increasing the complexity of the tree. A solution to this problem is to put redundancy in the tree, i.e., put several copies of the same subtree. These kinds of problems have been studied in great detail for hierarchical classifiers[3].

In the context of shape analysis, an analogous approach can be taken by specifying a strategy tree to be used in analyzing a shape. A strategy tree, T, is composed of a set of terminal nodes, $t_i$, a set of nonterminal nodes, $n_i$, and a root node r. The shape to be analyzed undergoes a sequence of tests on the path from the root to a terminal node, at which point it gets accepted or rejected with respect to the given grammar. The choice of the next set of constraints to apply and which symbol to produce depends on all the constraints generated from the shape primitives.

A strategy for interpreting the shape can be specified by a strategy tree which is composed of:

1. A tree skeleton or hierarchical ordering of the class labels (in our case, reject

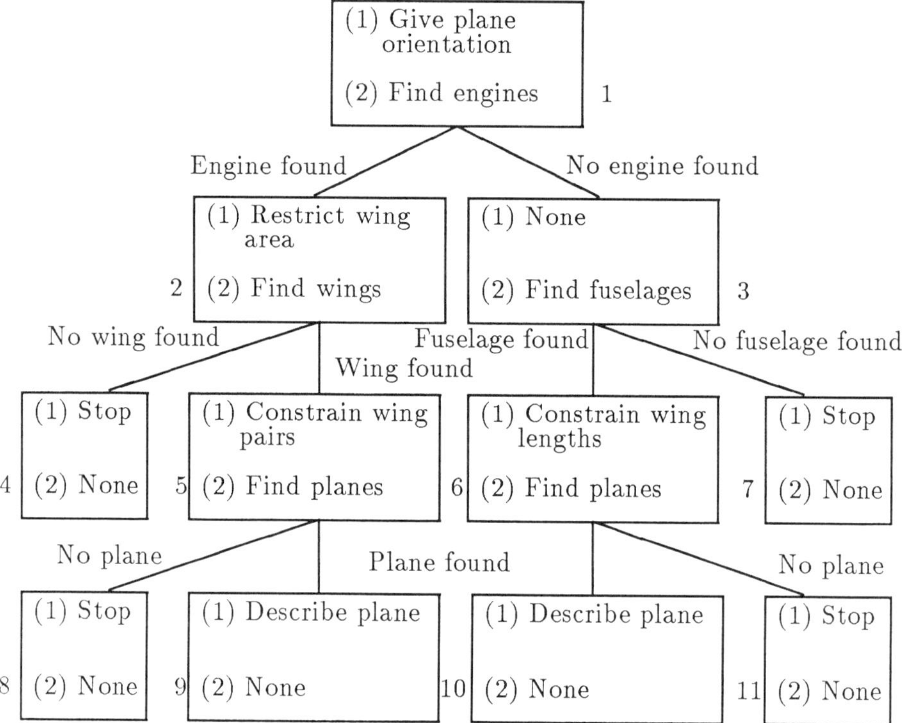


Figure 7.10: Strategy tree.

or accept the shape)

2. At each nonterminal node, a method of generating semantic constraints in the given context and an indication of how to apply the grammar to the remaining pieces of the shape

3. At each nonterminal node, a decision rule for choosing the next node in the tree

For example, suppose there is a grammar describing airplanes, then let T be as shown in Figure 7.10.

The strategy tree T describes a method for analyzing a shape. Node 1 gives the orientation of the main axis of any airplane. The semantic constraints will thus eliminate any engine side hypotheses that are not parallel to the given orientation. Next, all engine symbols are produced. If some engine symbols are produced, then Node 2 is the next node traversed. Node 2 constrains the search for wings by limiting search to the areas near engines. Next Node 2 produces wings. However, if Node 1 produced no engines, then Node 3 is traversed after Node 1. Node 3 specifies no new constraints and produces fuselages. If Node 2 produced no wings, Node 4 terminates the process. However, if Node 2 found some wings, then Node 5 makes sure that wings appear in pairs and then produces planes. If Node 3 produced some fuselages, then Node 6 specifies the relative lengths of wings to be computed (i.e., an allowed range of lengths for wing part hypotheses). Node 6 then produces plane symbols. If Node 3 found no fuselages, then Node 7 terminates the process. Likewise, if Node 5 or Node 6 produces a plane symbol, then Node 9 or Node 10, respectively, outputs a description of the plane; otherwise, the process terminates at Node 8 or Node 11, respectively.

The specification of a strategy tree can be done in a highly interactive manner which allows the user to give components of the tree in a detailed way. The tree can be input from the root node down, breadth-first. Each terminal node is either a command to terminate or a command to give the description of the shape. However, several items of information must be detailed for each nonterminal node. First, the number of sons must be input and their names determined. Next, a method for generating semantic constraint information (if any) and for applying the grammar is necessary. This method must be composed of operators defined for this purpose. Finally, a decision rule must be input, and this is accomplished by associating with each son a condition based on the result of applying the grammar.

The operators for generating the constraints should allow for manipulating the following:

1. Area to work on
2. Vocabulary symbols to find
3. Properties of vocabulary symbols, including
   - Length of certain axes of a vocabulary symbol
   - Orientation of certain axes of a vocabulary symbol

Once a strategy tree has been constructed, it can be applied to some set of segments to achieve a description of any shape for which a parse exists. The strategy tree uses HCP to find pieces of the shape, and having found the pieces, allows for focusing of the search on areas of interest. A strategy tree thus allows for user-directed search in applying the grammar to a shape.

Thus, we see that a method (strategy trees) for allowing more external control over the use of HCP has been defined, and it has been indicated how such a method can be applied to shape analysis. Programs exist which allow the user to define strategy trees; however, programs do not yet exist which implement strategy tree control over the application of HCP.

# Chapter 8

# Parallel Discrete Relaxation Algorithms

**This chapter was written in collaboration with Ashok Samal.**[1]

## 8.1 Introduction

One of the major drawbacks of using the discrete relaxation algorithms is their complexity. In general computer vision applications, speed is a major concern, and it is even more important in real-time applications. All the discrete relaxation algorithms (like most other algorithms) were designed for traditional von Neumann machines, with one processor and a single instruction stream, thus limiting the execution to one instruction per cycle. The recent widespread availability of multiprocessors has opened up avenues for improving the performance of these algorithms. (Another alternative is to use special-purpose hardware; see Chapter 9.)

Thus, we now need to study the algorithms from a different point of view. Some of the questions remain the same as before. For example, what are the computational complexities of these algorithms? What is the best an algorithm can do? However, there are several new issues that arise. How effectively can the processors be utilized? What is the greatest increase in speed that can be obtained? Should we modify the serial algorithms to fit a multiprocessing framework, or should we start fresh and develop new algorithms? Is it worthwhile to use multiprocessors for running these algorithms? We will attempt to answer some of these questions in this chapter.

The rest of the chapter is organized as follows. In Section 8.2 we describe the abstract machine model which is used to analyze the algorithms. We also discuss the notation to be used for the rest of the chapter. Section 8.3 discusses the various measures that are used to evaluate the performance of the parallel algorithms and their usefulness. In Section 8.4 we prove that the optimal complexity for arc consistency is $O(nm)$, where $n$ and $m$ are the number of nodes and number of labels per node, respectively. In Section 8.5 we analyze the parallel complexities of the various arc consistency algorithms. In Section 8.6 we compare the parallel algorithms and discuss the implications. In Section 8.7 some of the issues relating

[1] This chapter is a modified version of Reference 71.

implementation of these algorithms on a multiprocessor are explored. Finally, in Section 8.8 we briefly summarize the chapter and discuss future research.

## 8.2 Abstract Machine Model

In the case of uniprocessor machines, the computation model as defined by von Neumann is taken to be the standard. However, several models have been proposed to study the computation properties of multiprocessors (see[65] for a survey). The model we use for analysis of the algorithms in this chapter is the PRAM (Parallel Random Access Machine) model[72].

PRAM has an infinite number of processors, an unbounded amount of global memory, a set of input registers, and a finite program. The processors are capable of working both synchronously and asynchronously. Each processor resembles a von Neumann machine and has a local accumulator, a memory bank, and a program counter. However, the memory is unbounded. The global memory is shared by all the processors and any processor can read/write from/to any memory location at any time during execution. Both concurrent reads and concurrent writes are allowed in this model. However, if several processors try to write into the same location at the same time, any one (nondeterminate) of them will succeed.

The global memory is the only means of communication between the processors. This essentially sidesteps the issues concerning the communication network and overheads involved thereof. The instruction set of the machine consists of usual operations like LOAD, ADD, JUMP, etc. In addition a set of new instructions are added to incorporate the multiprocessing capabilities, e.g., fork processes, synchronization primitives, etc. The time to create a process (task) on a processor is assumed to be negligible.

Clearly the PRAM model can't be realized in hardware. However, there are several multiprocessors which come close to it; e.g., the class of shared memory multiprocessors. For the algorithms analyzed below, we assume that the number of processors that can be used is bounded by a polynomial in the size of the input. A solution to the consistent labeling problem can be found in constant time when an exponential number of processors is used. (Also, see references 7 and 70.)

Here we briefly describe the notation used in this chapter. The time taken by the algorithm to run on a single processor is denoted by $T_1$. $T_\infty$ is the time taken when infinitely many processors are available. It is also called the *unbounded parallel complexity*. Time taken under bounded parallelism [with a finite number ($k$) of processors] is denoted by $T_k$. $S_k$ is used for speedup obtained over the best sequential algorithm. $S_{\max}$ is the maximum possible speedup. While analyzing the algorithms, the unit of computation varies from algorithm to algorithm and will be clear from the context in most cases. In general $\alpha$ is used to denote the time taken to execute a unit-computation.

The algorithms are described in a Pascal-like language. Comments are given only when necessary. No attempt is made to make the algorithms syntactically correct Pascal functions/procedures. Several new constructs are used in the algorithms to describe the behavior in a multiprocessor framework. **P** and **V** are Dijkstra's[12] semaphore primitives for synchronization. To describe the synchronous mode of parallel computation, a parallel *for* construct is added. The construct

$$\| \; for \text{ i} := 1 to \text{ n} \; do \; \mathbf{f}(i);$$

creates $n$ tasks which execute concurrently (at least in theory), the $i$th task computing **f**($i$). The statement after || *for* is executed only after all the $n$ tasks are completed. The asynchronous computation is described by *fork* and *join* constructs.

A network of binary constraints is assumed. The sequential algorithms and their complexities are given in[52,53,58]. The analysis here will use some of their results. The terminology is also from the above references. Some of the most commonly used terms are explained below.

The network is assumed to consist of $n$ nodes and $e$ arcs. The number of labels per node is $m$. The labels are stored in a two-dimensional $n \times m$ array called **LabelSet**, and all the elements are initialized to 1. **LabelSet**$[i, j] = 0$ means that $j$ is not a viable label for node $i$. Also, it is assumed that the arcs are stored in an array of records, called ARCS. ARCS[$x$].start and ARCS[$x$].end refer to the start and end nodes of the arc $x$.

## 8.3 Performance Measurements

In the case of sequential algorithms, the two main concerns are their time and space complexities. These two remain very important issues for the parallel algorithms as well. However, these two figures alone don't portray a parallel algorithm accurately or completely.

For example, they don't indicate how much better the algorithm runs as compared to a sequential algorithm. Nor do they show how many processors are needed to achieve this. Several measures have been proposed to quantify many of these properties of parallel algorithms. We will define only a few of these and describe them briefly in this section. See References 70 and 77 for more of these measures.

### 8.3.1 Speedup

This is the most commonly used and probably the most important characteristic of parallel algorithms. The most common definition of *speedup* is the ratio of the time taken for a given algorithm to run on a uniprocessor and the time taken by it on a multiprocessor.

$$S_k = \frac{T_1}{T_k} \qquad S_{\max} = \frac{T_1}{T_\infty}$$

$S_k$ is the speedup obtained by using $k$ processors, $T_1$ and $T_k$ are the times taken by the algorithm to run on a single processor and on a multiprocessor with $k$ processors, respectively. $S_{\max}$ is the maximum speedup attainable by using the algorithm. In general it is a very good indicator of the success of a multiprocessor implementation.

There are several problems with this definition. First, it can be misleading. A high speedup does not necessarily mean a better algorithm. It is important to realize that this speedup number is only for the particular algorithm at hand. So it is possible that an algorithm $A$ has a higher speedup than an algorithm $B$, but actually runs slower on a multiprocessor. It might happen if the sequential version for $A$ is very slow to start with. Second, by transforming a sequential algorithm to run on a multiprocessor, we necessarily change its structure. Hence, it is not the same algorithm any more. So, it is not entirely accurate to say that algorithm $B$ is the parallel version of $A$. In fact, two sequential algorithms might be similar, but their parallel versions may have very different structures.

Keeping these points in mind, we define the speedup as the ratio of the time taken for the *optimal* sequential algorithm to the time taken for the algorithm to run on a multiprocessor. This definition takes care of the two points noted in the previous paragraph and portrays the algorithms better. If we didn't have a multiprocessor, surely we would go with the best sequential algorithm. So, any advantage one would obtain by using a multiprocessor should be relative to it. One of the problems with this definition is that many times the optimal algorithm is not known. In such cases it is still better to compare with the best known algorithm, than to compare with the "sequential version" of the algorithm. In our case however, we know that the optimal serial complexity of arc consistency is $O(em^2)$. So we compute speedup with respect to this, and $T_1$ for us means the time taken by the best sequential algorithm.

### 8.3.2 Efficiency

Efficiency is defined as the ratio of the speedup obtained and the total number of processors used by the algorithm.

$$\eta = \frac{S_k}{k}$$

This is relevant because it indicates how effectively the processors are used. Higher efficiency means better utilization and vice versa.

### 8.3.3 Maximum Effective Processors

Except for a few cases (combinatorial implosion [37,47,46]), speedup is less than the number of processors used. Ideally one would like the speedup to go up linearly (with slope 1) with the number of processors. However, depending on the problem size, at a certain point adding processors does not increase the performance of the algorithm. We call this the maximum effective processors, and it is denoted by $P_{\text{max}}$.

There are several other measures one could use to characterize the parallel algorithms, e.g., inherent parallelism [70,71], synchronization costs, etc. We restrict our self to the ones described above, in addition to the time and space complexities.

## 8.4 Lower Bound for Parallel Arc Consistency

In this section we derive the lower bound for any algorithm which enforces arc consistency in parallel. It is given by the following theorem:

**Theorem 8.1** *Any parallel algorithm enforcing arc consistency, in the worst case, must have $O(mn)$ sequential steps, where $n$ is the number of nodes and $m$ is the number of labels per node.*

**Proof**: Consider a network of $n$ nodes connected to form a cycle as shown in Figure 8.1. Let the label set of node $i$ be the set of numbers defined as: $\{x \mid x = jm + i, 1 \leq j \leq n\}$. The relation used here is the *greater-than* relation such that values at node $i + 1 \pmod{n}$ should be greater than values at node $i$. It is assumed that the network is node consistent before any algorithm is applied. In general the arcs of the graph may represent different types of constraints, and we may not be able to use any special technique to enforce global arc consistency. The only way to enforce the global arc consistency is to enforce consistency along each arc. If

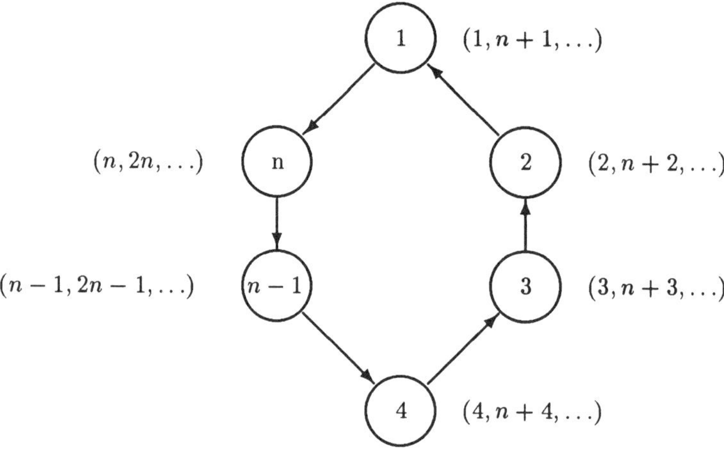


Figure 8.1: A network of constraints.

unlimited processors were available we could enforce consistency along all the arcs concurrently. In this problem instance, during the first time step only label 1 of node 1 is removed, since it doesn't get any support from node $n$. But all the labels at all other nodes have some support and hence remain intact. During the second time step only label 2 of node 2 gets deleted since, it was being supported by 1 at node 1 which was just deleted in the last time step. So, during the first $(m-1)n$ time steps, $(m-1)$ labels are removed from each node and the order of removal of labels from nodes is cyclic; $1, 2, \ldots, n$, $1, \ldots, n$, and so on. Then the label set of node 1 becomes empty, since it has no support from node $n$. After that the label sets of $2, 3, \ldots, n$, all become empty in 1 time unit, since the empty label set for one unit implies that there can be no globally consistent labeling. Thus the number of steps needed before any algorithm can terminate is $(m-1)n + 2$. So, parallel arc consistency is $\Omega(mn)$. In the above proof it is assumed that there are at least $e$ processors and they are working simultaneously on the arcs of the graph. If there are more processors the time taken in the worst case doesn't change. The theorem also holds if there are fewer processors.

## 8.5 Parallelism in Arc Consistency Algorithms

In this section we analyze the arc consistency algorithms along the measures defined in Section 8.3. Before we go into the details of the algorithms, it should be pointed out that all of them assume that the network is node consistent. We first give a parallel algorithm to enforce node consistency.

### 8.5.1 Parallel Node Consistency

A parallel node consistency algorithm is given below. $P_x$ is the unary predicate which has to be satisfied for node $x$.

**procedure**PNC(node);
**begin**

```
  ||for i := 1 to m do
     if ¬P_node(i) then LabelSet[node,i] := 0;
end

procedure Parallel-Node-Consistency();
begin
  ||for i := 1 to n do PNC(i);
end
```

**Algorithm PNC**. Parallel node consistency algorithm.

The serial complexity of the node consistency algorithm is $O(mn)$. If there are $mn$ processors, it is clear that node consistency can be achieved in constant time. Essentially there are $mn$ independent tasks and they can be done in parallel. Thus $T_\infty = O(1)$. If there are only $k$ ($< mn$) processors, then the $nm$ tasks can be divided among the $k$ processors and hence the time taken will be $O(mn/k)$.

$$S_{\max} = \frac{T_1}{T_\infty} = O(mn) \qquad S_k = \frac{T_1}{T_k} = O(k)$$
$$\eta = \frac{S_k}{k} = 1.0 \qquad P_{\max} = mn$$

### 8.5.2 Parallel REVISE

The *Revise* procedure is used by AC-1, AC-2, and AC-3 to enforce the consistency along a single arc $(i, j)$. It removes all labels in node $i$ which don't have any support from node $j$. We use a parallel "revise" in the same spirit. It is described below:

```
procedure Previse(i,j);
begin
  CHANGED := false;
  || for x := 1 to m do
  begin
     Support[x] := nil;
     || for y := 1 to m do
        if P_ij(x,y) then Support[x] := true;
     if Support[x] = nil then
     begin
        LabelSet[i,x] := 0;
        CHANGED := true;
     end
  end
end
```

**Algorithm Previse**. Parallel revise algorithm.

The algorithm is much like the serial REVISE procedure, except that independent tasks are done in parallel now. CHANGED is a globally shared variable used to indicate if *Previse* changed the label set of some node in the network. If $m^2$ processors are available, then the time taken by the *Previse* procedure is constant.

Hence, $T_\infty = O(1)$. If only $k(< m^2)$ processors are available the time taken by the algorithm is $O(m^2/k)$.

$$\begin{aligned} S_{\max} &= \frac{T_1}{T_\infty} = O(m^2) \quad S_k = \frac{T_1}{T_k} = O(k) \\ \eta &= \frac{S_k}{k} = 1.0 \quad P_{\max} = m^2 \end{aligned}$$

### 8.5.3 Parallel AC-1

Algorithm AC-1 is essentially a brute force algorithm. During each iteration arc consistency is enforced for every arc in some order. If there is any change in the label sets of any one of the nodes this process is repeated. The algorithm terminates when there is no change in the label sets of any of the nodes. In the parallel AC-1 (called PAC-1), all the arcs are checked for consistency in parallel. Although it is possible that the number of iterations taken by PAC-1 is different than AC-1, both will give the same (and correct) result. PAC-1 is sketched below:

```
procedure PAC-1();
begin
   Parallel-Node-Consistency();
   repeat
      CHANGED := false;
      || for x := 1 to e do
         Previse(ARCS[x].start, ARCS[x].end);
   until ¬ CHANGED;
end
```

**Algorithm PAC-1**. Parallel AC-1 algorithm.

As mentioned before CHANGED is a Boolean variable which is globally shared between all the processes. There is no need for its protection since only the effect matters, i.e., it does not matter which (or how many) labels have changed. Any change in the label set of any one node implies at least one more iteration.

The algorithm consists of both inherently serial and parallel parts. The repeat loop has to be done sequentially while the "revise" along each arc may be done in parallel. Also, as previously described, the "revise" process can itself be parallelized. The algorithm can potentially utilize up to $em^2$ processors: $m^2$ processors to do the *Previse* for each of the $e$ arcs simultaneously. So the || **for** loop in the PAC-1 procedure can be done in some constant time. The time taken by the algorithm is the number of times the **repeat** loop is executed, times some constant (say $\alpha$), for the part inside the loop. During each iteration (except the last) at least one label gets removed (otherwise the algorithm would terminate). Since the total number of labels is $nm$ the maximum number of iterations is also $nm$. If there are only $k(< m^2e)$ processors available, then the total parallelism inherent in the algorithm cannot be exploited, but a definite speedup can be obtained. During each repeat loop, there are $m^2e$ jobs which are executed concurrently. When there are only $k$ processors, only $k$ tasks can be done in parallel, thus the time taken to complete all these tasks is $O(m^2e/k)$.

$$T_\infty = \alpha nm = O(nm) \quad T_k = O(em^3n/k)$$

$$S_{\max} = \frac{T_1}{T_\infty} = O(em/n) \quad S_k = \frac{T_1}{T_k} = O(k/mn)$$
$$\eta = \frac{S_k}{k} = \frac{1}{mn} \quad P_{\max} = em^2$$

Thus, the expected speedup for the algorithm is linear in the number of processors up to $em^2$ processors. An important note which has been mentioned before is that the number of iterations taken by PAC-1 may be different from AC-1. This is due to the fact that in AC-1 the effect of any change mode by the *Revise* procedure by an arc will be sensed by the arcs which call it later, even during the same iteration. In PAC-1 this may not be the case since all arcs call *Previse* simultaneously. Another important observation is that PAC-1 is an optimal parallel algorithm, since it takes the minimum possible time $[O(mn)]$ in the worst case.

### 8.5.4 Parallel AC-2

```
procedure PAC-2();
begin
   Parallel-Node-Consistency();
   for i := 1 to n do
   begin
      || for j := 1 to e do
      begin
         s := ARCS[j].start; t := ARCS[j].end;
         Q[j] = Q'[j] = false;
         if (s = i) and (t < i) then Q[j] = true;
         if (t = i) and (s < i) then Q'[j] = true;
      end
      QEmpty := false;
      while ¬ QEmpty do
      begin
         QEmpty := true;
         || for j := 1 to e do
         begin
            bf if (¬ Q[j]) then continue;
            k := ARCS[j].start; m := ARCS[j].end;
            changed := Previse(k,m);
            if changed then UpdateQ'(k,m,i);
         end
         || for j := 1 tp e do
         begin
            Q[j] = Q'[j];
            Q'[j] = false;
            if Q[j] then QEmpty := false;
         end
      end
   end
end
```

**Algorithm PAC-2**. Parallel AC-2 algorithm.

AC-2 is similar to Waltz filtering in spirit. It makes the network consistent in a single pass through the nodes. Although its structure is more complex than AC-3 it is actually a special case of the latter. The parallel versions however are very different. PAC-2 is sketched below. Each node is processed in sequence. Processing in each iteration is done in parallel.

Q and Q′ are two Boolean arrays used to maintain the two queues. The procedure UpdateQ′$(k, m, i)$ adds to Q′ all the arcs $(p, k)$ such that $p \leq i$ and $p \neq m$. If $e$ processors are available, it can be done in unit time. No synchronization is necessary. The processor requirements for PAC-2 can be seen in the innermost || **for** loop. *Previse* needs $m^2$ processors to achieve maximum parallelism. So, the maximum processor utilization is $em^2$ or $e^2$, whichever is higher.

The *while* loop can in the worst case take $O(mn)$ time, because there are $mn$ total labels and in each iteration at least one label must be deleted; otherwise the algorithm would halt.

$$\begin{aligned} T_\infty &= O(mn^2) \quad T_k = O(em^3n^2/k) \\ S_{\max} &= \frac{T_1}{T_\infty} = O(em/n^2) \quad S_k = \frac{T_1}{T_k} = O(k/mn^2) \\ \eta &= \frac{S_k}{k} = \frac{1}{mn^2} \quad P_{\max} = em^2 \end{aligned}$$

### 8.5.5 Parallel AC-3

The structure of AC-3 is similar to AC-1. However, in AC-3 only the arcs associated with the nodes whose label sets have changed during the previous iteration are added to the pool of arcs to be checked for consistency. There are two ways one can parallelize AC-3. The first approach is to exploit parallelism in the REVISE procedure, but leave the structure of the algorithm untouched. The way the arcs are selected and added to the queue is the same as before. This version of the algorithm looks exactly the same as AC-3. This version [PAC-3-1] can in the worst case have $2e + m(2e - n)$ iterations [53]. Thus its time complexity is $O(me)$.

```
procedure PAC-3();
begin
   Parallel-Node-Consistency();
   List := All the arcs in Graph;
   repeat
      CHANGED := false;
      || for x := 1 to length(List) do
      begin
         Previse(ARCS[x].start, ARCS[x].end);
         if CHANGED then AddArcs(ARCS[x].start,ARCS[x].end);
      end
   until Empty(List);
end
```

**Algorithm PAC-3.** Parallel AC-3 algorithm.

Another way to parallelize AC-3 is to let all the arcs in the queue apply consistency checks simultaneously. During the next iteration the queue is constructed from the arcs associated with the nodes whose label sets have changed. PAC-1 is actually a special case of PAC-3, where the set of arcs chosen is the complete set of arcs for every iteration. The new algorithm (PAC-3) is given below:

AddArcs(j,m) adds to List all the arcs $(i, j)$ in the network such that $i \neq j$ and $i \neq m$. Clearly, in the worst case the number of times the *repeat* loop is performed is $nm$ for the same reasons as for PAC-1. The maximum number of arcs in the List is $e$ since no arc is duplicated. So the maximum number of processors needed to do *Previse* in parallel is $e$. Also the AddArcs can be done in constant time if there are $n^2$ processors available for doing this task. This can be accomplished by using a globally shared array of arcs with a field to indicate the inclusion/exclusion of an arc in the queue. A similar argument holds for initialization of List.

Thus PAC-3 in the worst case is $O(mn)$. Also, using similar arguments as in PAC-1, if there are only $k$ processors, the time taken by the algorithm is $O(em^3n/k)$. Thus:

$$
\begin{aligned}
T_\infty &= \alpha nm = O(nm) \qquad T_k = O(em^3n/k) \\
S_{\max} &= \frac{T_1}{T_\infty} = O(em/n) \qquad S_k = \frac{T_1}{T_k} = O(k/mn) \\
\eta &= \frac{S_k}{k} = \frac{1}{mn} \qquad P_{\max} = em^2
\end{aligned}
$$

Although the parallel complexity of PAC-3 is the same as PAC-1, it has some drawbacks. First, the cost of building the list of arcs is not negligible. Although in general PAC-3 uses fewer processors than PAC-1, in the worst case it may end up using the same number of processors to get maximum parallelism. Since the number of processors needed keeps fluctuating, it may not be convenient to use the idle processors for some other purpose.

### 8.5.6 Parallel AC-4

In AC-4 several data structures are used to reduce the time complexity. The algorithm consists of three stages: building the appropriate data structures: M, S, and counter, initialization of List, and pruning the inconsistent labels. We sketch the parallel counterparts for each of the stages separately for ease of understanding.

**Construction of Data Structures**

```
procedure BuildDS();
begin
   || for x := 1 to e do
   begin
      i := ARCS[x].start;
      j := ARCS[x].end;
      || for y := 1 to m do
      begin
         total[x,y] := 0;
         b := LabelSet[i,y];
         || for z := 1 to m do
         begin
            c := LabelSet[j,z];
            if R(i,b,j,c) then
            begin
               P(Sem[x,y]);
```

```
            total[x,y] := total[x,y] + 1;
            V(Sem[x,y]);
            Append(S_jc, (i,b));
         end
      end
      if (total[x,y] = 0) then
      begin
         M[i,b] := 1;
         LabelSet[i,b] := 0;
      end;
      else counter[(i,j),b] := total[x,y];
   end
 end
end
```

**Algorithm PAC-4.** Step 1 of parallel AC-4 algorithm.

**Sem** is an array of semaphores used to synchronize the update of **total** which is an array to keep count of supports. P and V are Dijkstra's primitives for synchronization. This part can utilize up to $em^2$ processors effectively. They compute and update the data structures in parallel. The bottleneck here is in updating **total**. In the worst case all the updates of **total** are serialized and hence take $O(m)$ steps. It should be pointed out here that, appending and maintaining S can be done in unit time, by using a four-dimensional bit array. Thus $T_\infty = O(m)$ and $P_{\max} = em^2$.

```
procedure InitList();
begin
  || for i := 1 to n do
     || for b := 1 to m do
     begin
        if (M[i,b] = 1) then List[i,b] := 1;
     end
end;
```

**Algorithm PAC-4.** Initialization of **List**.

Here $mn$ processors concurrently update the elements of **List**. Since updating can be done in unit time, the procedure can be done in constant time. Thus $T_\infty = O(1)$ and $P_{\max} = O(mn)$. One could eliminate this procedure by using the **M** array directly. However in actual implementation one might prefer to separate the two.

We need P and V operations in the following scenario. Suppose there are two labels, say $c1$ and $c2$, in node $j$ which support the same label, say $b$, at node $i$. If both of them are in the **List** during the same iteration, then both might attempt to decrement **counter**[(i,j),b] at the same time.

Clearly the maximum number of elements in **List** is $mn$. Also, once a pair $(i, b)$ is removed from **List** it is never processed again. In the worst case, a single element is processed at each iteration of the *while* loop. Hence the maximum number of iterations performed by the *while* loop is $mn$. Due to the synchronization in the innermost loop, in the worst case it can take $O(m)$ time. However, if there are more than one (say $x$) identical counter updates (with same $i, j, b$) it means there are $x$

labels in node $j$ which support label $b$ at node $i$. It also implies that all of these $x$ labels are deleted, since they are in **List**. Thus, there are $x$ label-node pairs $(j, c)$ which get processed during the same iteration, which means there can now be at most $mn - x$ iterations. Thus the overall complexity remains $O(mn)$.

```
procedure PruneLabels();
begin
   while ¬ Empty(List) do
      || for x := 1 to Length(List) do
      begin
         (j,c) := List[x];
         List[j,c] = nil;
         || for y := 1 to Length(S_jc) do
         begin
            (i,b) := S_jc[y] ;
            P(Sem[(i,j),b]);
            counter[(i,j),b] := counter[(i,j),b] - 1;
            V(Sem[(i,j),b]);
            if (counter[(i,j),b] = 0) and (M[i,b] = 0) then
            begin
               Append(List, (i,b));
               M[i,b] := 1; LabelSet[i,b] := 0;
            end
         end
      end
   end while
end;
```

**Algorithm PAC-4.** Pruning inconsistent labels.

The maximum length of $\mathbf{S}_{jc}$ is $m(n-1)$, since label $c$ at node $j$ may support all labels at the other $n-1$ nodes. Then $T_\infty = O(mn)$ and $P_{\max} = m^2n^2$. If there were only $k$ processors, then the time required is $O(m^3n^3/k)$.

The parallel AC-4 (PAC-4) consists of calls to BuildDS, InitList, and PruneLabels sequentially. Thus, the time taken by PAC-4 is the sum of the three procedures and the processor requirement is the maximum of the three. The complexity of AC-4 is $O(em^2)$[58].

$$
\begin{aligned}
T_\infty &= O(m) + O(1) + O(mn) = O(mn) \\
T_k &= O(m^3n^3/k) \qquad T_1 = O(em^2) \\
S_{\max} &= \frac{T_1}{T_\infty} = O(em/n) \qquad S_k = \frac{T_1}{T_k} = O(ke/mn^3) \\
\eta &= \frac{S_k}{k} = \frac{e}{mn^3} \\
P_{\max} &= Max\{O(em^2), O(mn), O(m^2n^2)\} = O(m^2n^2)
\end{aligned}
$$

## 8.6 Discussion

In this section the parallel algorithms for enforcing arc consistency are compared with respect to the measures defined in Section 8.3. The most important of them is the time complexity. As has been shown, PAC-1, PAC-3, and PAC-4 are all $O(mn)$ algorithms. Hence, they are all optimal parallel algorithms. Another interesting observation is that although AC-2 is a special case of AC-3, their parallel counterparts are very different and have different complexities. It is because the structure of AC-2 is not very suitable for parallel processing.

The processor requirements of the three algorithms are also different. PAC-1 always needs $O(em^2)$ processors to exploit all the parallelism in the algorithm. Although both PAC-2 and PAC-3 need less processors on the average, in the worst case they still need $O(em^2)$ processors. Also, due to the uncertain need for processors from iteration to iteration it is not easy to utilize the idle processors in any effective way. PAC-4 needs $O(m^2n^2)$ processors. Thus, in general PAC-4 needs more processors. However, if the graph is complete (or close to it) then $e$ is $O(n^2)$. In this case the processor requirements of the four PACs are identical. Thus, the worst case processor requirement to exploit maximum parallelism is the same for all the parallel algorithms.

The speedup curves for all the algorithms are linear in the number of processors, up to the point of $P_{\max}$. Also the slopes of their curves are roughly the same, except for PAC-2. This is due to the fact that all except PAC-2 are optimal. It should be also remembered that the speedups are with respect to the best sequential algorithm (AC-4). Accordingly, the maximum speedups obtainable for these algorithms are the same except for PAC-2, which has the least speedup.

Since all three algorithms are optimal, the obvious question is which algorithm performs best on the average? How much time is taken up by the communication and synchronization overhead? None of the algorithms have any explicit communication overhead; all the communication is done through shared memory. All the algorithms need synchronization steps at the end of parallel *for* iterations. PAC-4 needs further synchronization while building the data structures, as explained before. PAC-3 on the average does less computation during each iteration, but it has additional costs of maintaining some lists. This cost can be minimized by trading off space. The structure PAC-4 is totally different from PAC-1 and PAC-3, and hence it is difficult to compare their average execution times.

## 8.7 Implementation Issues

The analysis, discussions, and results so far are theoretical. We have computed speedup and efficiency theoretically without actually running the algorithms on a multiprocessor. Measures computed this way are in some sense less *realistic* since we don't take into account the cost of communication, process creation overhead, etc. However, they are necessary because they give the estimates of these numbers under ideal (but probably unrealistic) circumstances. Experimental measures on the other hand have the deficiency that they are too specific, i.e., the numbers obtained by running an algorithm in a particular architecture don't generalize to others. Thus, they don't indicate the maximum speedup possible in any kind of architecture.

So, for a complete understanding of an algorithm in a multiprocessor environment, it is necessary to analyze it theoretically, as well as to implement it on a multiprocessor. It will also answer questions pertaining to which algorithm performs the best on the average. We briefly discuss some important implementation issues.

One measure which can be used here, is the *equivalence point*. It is defined as the number of processors required to match the performance of the best sequential algorithm. Although this can be used in theoretical analysis as well, it makes more sense in actual implementation.

First is the availability of processors. The processor requirements of these algorithms is very large, $O(em^2)$. Even for small problems, e.g., say $e = 20$, $m = 10$, the number of processors needed is 2000. There are multiprocessors which have this many processors, e.g., the Connection Machine [38] and MPP [64], but they are still not very common. Also, the granularity of computation is very fine. The unit of computation is a comparison of a pair of labels (in *Previse*). If relatively fewer processors are available, one may still get maximum benefit by increasing the granularity of computation. It is conceivable to delegate a single processor to do the REVISE along an arc, instead of doing it on $m^2$ processors. Clearly, how fine or coarse the granularity should be is determined by the number of processors available. The main observation is that even though these algorithms can use a large number of processors effectively, they can still be very efficient with smaller multiprocessors.

Another side issue is the effective utilization of the processors. Some of the algorithms do not use all the processors all the time. Their requirements change from one iteration to another. It is difficult to utilize those processors effectively during the period of inactivity, but one might take advantage of them to do some other useful tasks, perhaps in advance (speculative computation).

Another important consideration is memory contention. In our abstract machine, the global memory is truly global, i.e., the time taken by each to access it is the same for all processors. It is, however, not very realistic. Usually, each processor has some amount of local memory and if other processors want to access it, they have to go through the interconnection network which connects them. The amount of time taken to access the remote memory is typically larger than the time to access memory that is local. For example, in the BBN Butterfly parallel processor[44], the ratio of the access times for remote memory to local memory is about 15 to 1. Also, this time may either be uniform or may depend on the distance between the two processors trying to communicate. Thus, the interconnection network plays a very important role in the performance of the parallel algorithms.

This has one major implication in our case. We have data structures which have to be globally shared, e.g., the label sets of the nodes. If they are stored in one place, then the processors try to access that memory at the same time, which leads to severe hot spots and results in tremendous degradation in performance. Also, if they are distributed so as to minimize the memory contention, every processor must be made aware of the location of various data segments. However, it might not be possible to distribute data all the time. For example, we use a single variable (CHANGED) to detect the change made while doing the REVISE in parallel. Clearly, every process that deletes a label must set this variable to true. Although there is no lock needed while updating this variable, and the operation takes only a very short time, there might still be memory contention. We also have several places where we need explicit locking of memory in PAC-4. Depending on the circumstances it might also lead to hot spots.

Another issue is space requirements, although with the cost of memory chips going down rather rapidly, it is not a major concern. Also, the space requirements of the algorithms are not very large. The extra space needed for synchronization is comparatively small. So, one may try various tricks to utilize the extra space. For example, one may keep duplicate copies of data to avoid contention and periodically update them to maintain consistency.

Since there are several kinds of multiprocessors, it is important to consider what kind of architectures these algorithms are best suited for. Clearly, shared memory is a big help, since these algorithms use it rather extensively. Another parameter is the choice of SIMD or MIMD machines. The way algorithms are described show that they can be easily done on a SIMD machine. || *for* is essentially a SIMD construct. However, they can with equal ease be implemented on a MIMD machine.

## 8.8 Conclusions

It has been shown that parallel arc consistency is $\Omega(mn)$. Parallel algorithms for node consistency and several arc consistency algorithms are described and are analyzed for their complexities and other measures. Although the serial arc consistency algorithms have different complexities, the parallel algorithms (PAC-1, PAC-3, PAC-4) have the same time complexity. They all have the complexity of $O(mn)$, which means that they are all optimal (parallel) algorithms. It is not immediately clear which has the best average execution time. The algorithms also have similar processor requirements.

The speedup is computed with respect to the best sequential algorithm because it portrays the parallel algorithms better. Clearly the work described here is theoretical. To understand the performance of the algorithms better, they need to be implemented in multiprocessors. Several implementation issues are also discussed. Results of running these algorithms on a BBN Butterfly parallel processor are given in References 70 and 71.

In solving consistent labeling problems, node, arc, and path consistency can be used to cut down the search space. However, due to excessive computation needs of path consistency algorithms they are seldom used. But it might be worthwhile to examine them again in a parallel computation model and see if they can be useful.

# Chapter 9

# VLSI Architectures for Discrete Relaxation

**This chapter was written in collaboration with Jun Gu and Wei Wang.[1]**

## 9.1 Introduction

Discrete relaxation techniques have proven useful in solving a wide range of problems in digital signal and digital image processing, artificial intelligence, operations research, machine vision, and neural network computing. Since the introduction of the discrete relaxation technique in the mid-seventies, many DRA applications have been found, and significant progress has been made to develop new DRA algorithms[35,58,67].

On the other hand, people have also tried to design efficient, parallel hardware algorithms and computer architectures to support fast, real-time solutions to the Consistent Labeling Problem (CLP). Unfortunately, due to high computational cost including time complexity, data routing and communication complexities, simultaneous data pattern generation complexity, and space complexity, the most straightforward hardware design, DRA1, suffers from $O(n^2m^3)$ time complexity and $O(n^2m^2)$ space complexity[48], for an $n$-object and $m$-label DRA problem[48]. Past research in this aspect reached an impasse, and appeared only in a virtual software simulator format[55].

This chapter presents some recent work in developing special-purpose DRA VLSI computer architectures. Certainly, any parallel DRA algorithm can be easily mapped onto to any general-purpose multiprocessor computer system without much difficulty. However, several problems arise when building a special-purpose, cheap, and fast DRA architecture. First, except for a few cases, speedup on a multiprocessor is less than the number of processors used. Ideally one would like the speedup to go up linearly (with slope 1) with the number of processors. For many CLP applications, even when 30 to 50 processors are used, the execution (CPU) time, compared with the execution on a sequential machine, still remains on the same order. Second, after a certain saturation point, adding processors does not increase the speedup. Third, the overhead incurred in a multiprocessor system, such as creation of processes, synchronization, and remote memory access, is very expensive.

[1]This chapter is a modified version of Reference 27.

Fourth, for real-time computing, the hardware cost for a general-purpose multiprocessor machine is usually very expensive. Finally, most such machines require a lot of physical space and care.

Our primitive strategies for designing DRA architectures can best be described by introducing the parallel DRA2 and DRA3 systems. By manipulating DRA into a parallel computational tree, a parallel hardware DRA2 algorithm is formulated. This algorithm is implemented in the DRA2 architecture using a multi-dimensional tree-root pipelining scheme on an $n$ by $m$ SIMD multiprocessor array. Its time complexity is reduced to $O(nm)$, while the space complexity is reduced by a factor of 2. For certain relaxation processing, the space complexity can even be decreased to $O(nm)$. Furthermore, a theoretical analysis of the DRA data pattern was performed. As a result, a highly concurrent, dynamically structured DRA3 algorithm was created. A technique for dynamically configuring an architectural wavefront is used which generates the DRA data pattern simultaneously and leads to an $O(n)$ time highly configurable DRA3 architecture.

The hardware implementation issues of three discrete relaxation architectures are addressed. These three architectures were designed for the same 8-object 8-label DRA problem. The first two systems were implemented using PPL (Path Programmable Logic)[76] at the University of Utah. The DRA1 chip requires two 4K memory blocks and maximum execution time of several minutes in a 3-$\mu$m NMOS process, which makes such a hardware implementation infeasible[48,90]. The DRA2 design eliminates excessive memory requirements and performs the DRA computation, in the worst case, in microseconds. This chip was fabricated using a 3-$\mu$m NMOS process by MOSIS[60,90,89]. DRA3 is fabricated using a 1.2-$\mu$m CMOS technology. The clock speed of DRA3 is expected to be over, at least, 35 MHz[27].

This chapter is organized as follows. In Section 9.2 we define the DRA hardware implementation problem. Complexity analyses and our first design of the DRA1 chip are presented in Section 9.3. In Section 9.4, we concentrate on the design of the parallel DRA computational tree and a multi-dimensional tree-root pipelining scheme, and its corresponding $O(nm)$ time SIMD multiprocessor architecture, DRA2. Finally, it is shown that we have adopted a dynamically configurable, highly parallel data pattern routing and generation scheme on the DRA3 switch lattice, so that an $O(n)$ time algorithm-configured architectural wavefront system, DRA3, is produced. Also, in this chapter, we describe research ideas and the architectural concepts for these DRA architectures as concisely as possible. For detailed references see References 22, 27, 60, 91 and 92.

## 9.2 DRA Hardware Implementation

### 9.2.1 Assumptions and Statements

To concentrate our attention on tackling the major design complexities without losing generality for developing various DRA architectures, the following assumptions are adopted in our implementations:

**Assumption 9.1** *The number of objects and labels is assumed to be equal, i.e., $n = m$. Since the extensibility for different numbers of objects and labels has first been considered to be an important factor in system design, DRA2 and DRA3 can be specified for an arbitrary number of objects and labels as long as chip size permits.*

**Assumption 9.2** *For useful DRA applications, we have chosen the minimal value of $n$ and $m$ to be 8. This is reasonable, for example, in image analysis applications.*

**Assumption 9.3** *Contrary to the assumption in Reference 55 that the input data are always ready during the DRA computation, we assume that both data load-in time and their hardware support cannot be ignored in a VLSI DRA computation. This assumption provides a greater challenge for an advanced DRA system implementation.*

**Assumption 9.4** *The time for binary operation is generally derived from one gate delay of the simple combinational logic circuits, which is, in our DRA designs, always less than one global clock cycle. Similarly, it is common that one clock cycle is usually less than one Read/Write memory cycle. That is, we have that*

$$t_{\text{gate delay}} < t_{\text{clock cycle}} \leq t_{\text{memory R/W cycle}} \tag{9.1}$$

Whenever we analyze the time complexity, the following three statements hold:

**Statement 9.1** *Since the fact that the time complexity for each DRA computing iteration is the same, theoretically, we estimate the time complexity only within one iteration of computation.*

**Statement 9.2** *As for the order of iterations, which is solely determined by the convergence property of the algorithm and problem feature, we give its estimate in terms of the corresponding VLSI fabrication technology.*

**Statement 9.3** *We also follow the traditional notation for time complexity analysis for the DRA system, i.e., we describe it according to the number of objects, $n$, and the number of labels, $m$, and not the bit number of input data. For example, following statement 1, it takes $O(n)$ time to compute the DRA solution in the DRA3 architecture, where $n$ is the number of objects, whereas we have actually loaded in $n$ by $m$ bits of input data for each computation.*

### 9.2.2 DRA Hardware Implementation Problems

**Definition 9.1** The DRA2 hardware implementation *problem has been defined as finding the* consistent labeling matrix **L***:*

$$\mathbf{L} \quad = \quad (\mathbf{L}_1, \mathbf{L}_2, \ldots, \mathbf{L}_i, \ldots, \mathbf{L}_n)^t \tag{9.2}$$

*for the given* region coloring problem, *provided by the* initial labeling matrix*:*

$$\Lambda \quad = \quad (\Lambda_1, \Lambda_2, \ldots, \Lambda_i, \ldots \Lambda_n)^t \tag{9.3}$$

*and the* object-label pairs' compatibility matrixes $\mathbf{C}_{ij}$ *for every $i$ and $j (i, j = 1, 2, \ldots, n)$.*

**Definition 9.2** The DRA3 hardware implementation *problem has been defined as finding the consistent labeling matrix* **L***:*

$$\mathbf{L} \quad = \quad (L_1, L_2, \ldots, L_i, \ldots, L_n)^t \tag{9.4}$$

*for any given* world model, *provided by the* initial labeling matrix*:*

$$\Lambda \quad = \quad (\Lambda_1, \Lambda_2, \ldots, \Lambda_i, \ldots \Lambda_n)^t \tag{9.5}$$

*and the* object-label pairs' compatibility matrixes $\mathbf{C}_{ij}$ *for every $i$ and $j (i, j = 1, 2, ..., n)$.*

The above definitions represent our strategy for developing DRA architectures. The DRA2 system is aimed at first reducing both the space and time complexities for a particular DRA application problem, i.e., the region coloring problem. The second one, the DRA3 design, generalizes DRA2 for an arbitrary DRA problem and leads to a faster and more general-purpose computer architecture.

## 9.3 DRA1 Design and Complexity Analysis

The entire DRA computation consists of mainly two parts. The first part generates the $n^2(m \times m)$ compatibility matrixes for object pairs $u_i$ and $u_j$; i.e., computing each matrix $\Lambda_{ij}(k,p)(i,j = 1,...,n; k,p = 1,...,m)$ from the most recently evaluated (or originally given at the first iteration) labelings. The second part is an iteration process for the relaxation of the current object pairs' constraints among all different objects to produce new labelings; i.e., computing each new labeling $l_{ik}(i = 1,...,n; k = 1,...,m)$, provided by the currently known object constraint matrixes $\Lambda_{ij}(k,p)$. Reviewing the DRA formulation and the computational procedure for the region coloring example in Chapter 4 reveals the complexity issues met in the DRA1 system design.

**Time complexity** At least $n \times m$ memory Read/Write operations must be performed for computing an $n \times m$ matrix of $l_{jp}^{(n-1)} * \Lambda_{ij}(k,p)$ elements. Because there are $n \times m$ iteration equations (for computing new labels $l_{ik}, i = 1,\ldots,n; k = 1,\ldots,m$) which need to be evaluated, the sequential computation time would be $O(n^2m^2)$. In the DRA1 design, in addition to using the sequential evaluation strategy, another $O(m)$ time is spent on the complicated nested control loop for memory Read/Write operations during each iteration evaluation. Thus the entire computation time within each iteration costs $O(n^2m^3)$. There are other binary computations which must be taken into account. According to assumption 4, their contributions compared to the R/W operations of memory are ignored.

**Space Complexity** During each iteration of the DRA computation, several intermediate arguments (initial arguments at the first iteration) take certain amounts of space to be stored. The initial label matrix $\Lambda$ ($\mathbf{L}^{(0)}$ matrix) and the next iteration result of $\Lambda$ each takes $O(n^2)$ space. The matrixes $\Lambda_i \times \Lambda_j^t$ and the label pairs' compatibility matrixes $\mathbf{C}_{ij}(k,p)$ as well as the resulting object pairs' compatibility matrixes $\Lambda_{ij}(k,p)(i,j = 1,\ldots,n; k,p = 1,\ldots,m)$ each takes $O(n^2m^2)$ space. The total space complexity is:

$$O(2n^2 + 3n^2m^2) \text{ which is } O(n^2m^2) \tag{9.6}$$

**Data routing complexity** It is meaningful to consider the routing complexity only if a parallel DRA architecture may possibly exist. We discuss this issue in Sections 9.4 and 9.5 when developing parallel DRA architectures. In fact, the data routing complexity during each iteration is over $O(n^2m^2)$. For an 8-object 8-label DRA parallel system, there are at least 8K data being routed or communicated during each iteration.

The DRA1 architecture, unfortunately, serially computes each intermediate element. The interphase computation strategy imbedded in this design is purely *I/O-bounded.* Assume $t_{\text{read}} \approx t_{\text{write}} \approx 500$ns for an NMOS process and the time complexity is $O(n^2m^3)$ of each iteration. Multiplying the possible maximum iterations, which is on the order of $O(nm)$ and is determined by the feature of the computational model, the worst case execution time is several minutes. For practical applications, the number of objects could be 8, 16 or 32, the corresponding memory requirements for these different cases are 8K, 32K and 128K, respectively. As shown in the DRA1 design, this has greatly added to the circuit size which is a bottleneck in the DRA1 system design when $n$ becomes large.

## 9.4 The Parallel DRA2 System

In the analyses above, most of the computing time was spent on the interphase R/W operations for matrixes $\Lambda_{ij}(k,p)$; moreover, one half of the $O(n^2m^2)$ space is taken for the generation of matrixes $\Lambda_{ij}(k,p)$, which blocks the DRA1 VLSI computation. In this section we describe a DRA2 architecture which is completely independent of the $\Lambda_{ij}(k,p)$ evaluation.

### 9.4.1 A Parallel Tree-Structured Reformulation for the DRA2 Algorithm

It should be clear that any attempt to speed up an I/O-bound computation must rely on an increase in the memory bandwidth. Speeding up a compute-bound computation, however, may frequently come from the concurrent use of many processing elements[50]. The degree of parallelism in a special-purpose computer system is largely determined by the underlying algorithm. In order to solve the complexity arising in the DRA1 design, the following three steps have been taken to design a hardware algorithm that supports a high degree of concurrency in DRA computation.

#### Constructing the Parallel Computation Tree

When more effort is spent analyzing equation 4.4, we see that element $\Lambda_{ij}(k,p)$ can be decomposed as:

$$\Lambda_{ij}^{(n-1)}(k,p) = l_{ik}^{(0)} * l_{jp}^{(0)} * \mathbf{C}_{ij}(k,p) \tag{9.7}$$

which can form a leaf node as shown in Figure 9.1: which can be hierarchically formed as a tree-like structure with each level imbedded in the parallel computation for their leaves' operands as shown in Figure 9.3.

#### Speeding Up Iteration

Once a DRA problem in processing is identified as convergent, its current computing labeling reaches its final consistent labeling asymptotically in such a way that the error vectors, i.e., the difference vectors between these two labelings, monotonically decrease. Therefore, the node computation in Figure 9-1 can be speeded up by replacing the initial given elements $l_{ik}^{(0)}$ and $l_{jp}^{(0)}$ with the most recently $(n-1)$th iterated results $l_{ik}^{(n-1)}$ and $l_{jp}^{(n-1)}$. The modified computation is composed of a new leaf node as shown in Figure 9.2. The computation tree for $l_{ik}^{(n)}$ is formed as shown in Figure 9.2. In Section 9.4.2 it is shown that the bottommost leaves' operands have been associated with a data pipelining channel, completing a multiple parallel multi-dimensional tree-root pipelining scheme which supports a highly concurrent

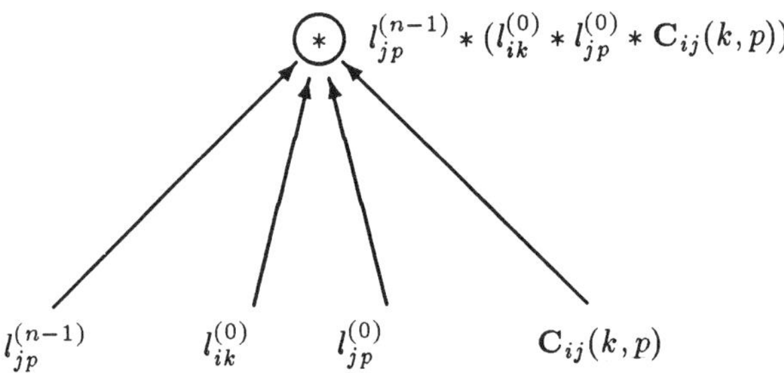


Figure 9.1: Leaf node for computing $l_{jp} * \Lambda_{ij}(k,p)$.

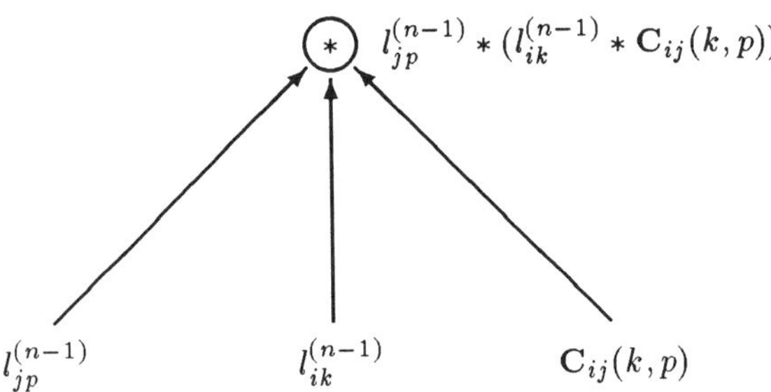


Figure 9.2: Modified leaf node computation for $l_{jp} * \Lambda_{ij}(k,p)$.

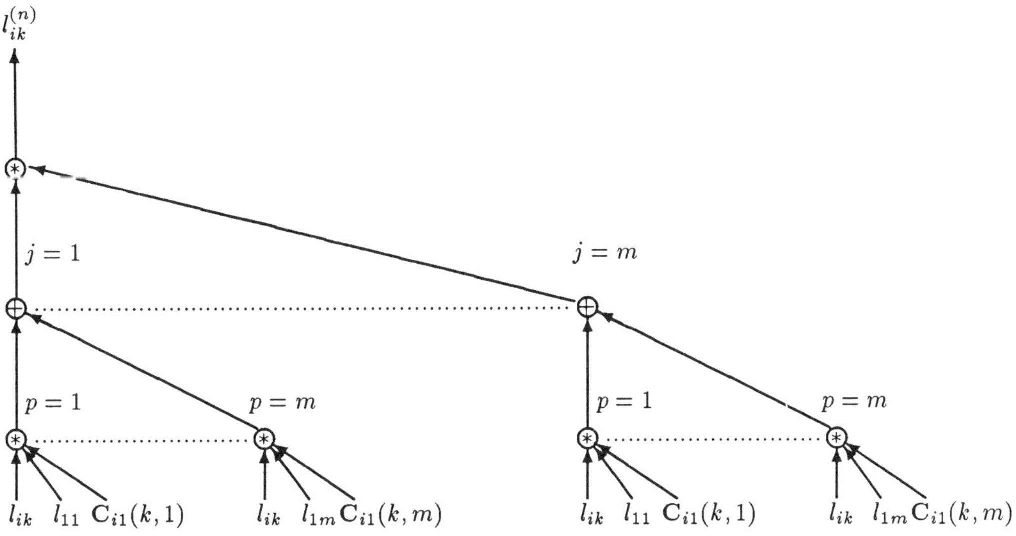


Figure 9.3: A parallel tree for computing $n$th $l_{ik}$.

DRA computation. For convenient notation in the rest of the chapter, we classify these three kinds of bottommost leaves' operands of Figure 9.3 in the following definition.

**Definition 9.3** *(1) We define the labeling elements of the $l_{jp}$'s ($j = 1, \ldots, n; p = 1, ..., m$), which are inside the pipelining channel, to be the* l_pipe; *and those on the tree root to be the* l_root. *Both* l_pipe *and* l_root *have the same logic values except they are topologically separated in the system layout. (2) Signal $l_{ik}$ passes through all roots horizontally through each row array of $k$; it is defined as the broadcasting signal $b_k$. (3) The $\mathbf{C}_{ij}(k,p)$ matrix elements, which are systematically distributed along each tree root, are defined as the* $\mathbf{C}_{ij}(k,p)$-Pattern.

### Introducing a Time Dimension in the Computation

To compute an $n$-object $m$-label DRA problem, a total of $nm$ $l_{ik}$'s need to be evaluated. This means at least 64 computation trees as shown in Figure 9.3 need to be built inside the circuit for our $n = m = 8$ case, which greatly increases the circuit size. To minimize this problem, each operand at the bottom of the tree has been constructed with a time dimension. As time progresses, different $l_{ik}$'s ($i = 1, \ldots, n; k = 1, \ldots, m$) can be generated. These time-varying characteristics can be best described by the following *spatial-temporal (ST) index equations*:

$$i \leftarrow i + t \bmod n \tag{9.8}$$

$$j \leftarrow j + t \bmod n \tag{9.9}$$

denoted by $A \leftarrow B$ or we say that $A$ is generated by $B$. We call $i$, or $i+t$, and $j$, or $j+t$, the spatial-temporal indexes since they characterize the indexes in the recursive formulas. They constitute the theoretical basis for the recursive systolic computation and interleaved processing in the DRA2 and DRA3 systems, while the basic DRA-PE cell realizes the computation in the formula. It is also of interest to see that in the DRA2 system, the ST indexes generate the dynamic forward DRA computational wavefront on a statically configured DRA architecture; while in the DRA3 architecture, it forms the dynamically configurable DRA3 architectural wavefront for a virtually static computation.

## 9.4.2 Implementation Issues for the DRA2 Chip

### System Architecture and Block Diagram

The block diagram of the DRA2 circuit is illustrated in Figure 9.4. The chip consists of four functional blocks.

1. Compatibility Matrix Registers (CMR): $C_{ij}$ registers are a set of eight 8-bit shift registers in the leftmost part of the circuit; they are used for storing each $\mathbf{C}_{ij}$ matrix. Another set of $\mathbf{C}_{ii}$ registers in the rightmost part of the circuit are for storing $\mathbf{C}_{ii}$.

2. $8 \times 8$ Multiprocessor SIMD Array (MSA): The MSA is composed of 8 by 8 simple and regular cells. They are predefined to map the highly parallel computation DRA2 algorithm of Figure 9.3 onto silicon. A number of parallel horizontal and vertical communication wires are designed around the four edges of the cells to make use of higher degrees of parallelism in the computation.

3. L-Matrix Shift Register (LSR): It is used for (1) the input and output data paths for the original and final labeling matrixes; (2) the pipelining channel for multi-dimensional tree-root operand broadcasting and pipelining, forming a recursive DRA computational wavefront; and (3) performing temporarily the data storing and updating.

4. Control Module (CM): This module includes four units. An 8-bit comparator is located on top of the first 8-bit shift register of the LSR to sense the equality between the $n$th output vector $\mathbf{L}_i^{(n)}$ of the MSA array and the corresponding $(n-1)$th row vector $\mathbf{L}_i^{(n-1)}$ inside the LSR. A timer is served as both the systole pacer and tagged-bit signal generator for iteration control. An 8-bit state register is used for collecting comparison results from the comparator and monitoring iteration states. Finally a finite state machine (FSM) is built for performing a self-timed synchronization among these functional blocks and host computer.

This diagram of four functional blocks also serves as the PPL layout floor plan for efficient layout (in subsection 9.4.4).

### Multiprocessor SIMD Array and the Cell Design

The basic principle of the multiprocessor SIMD architecture for DRA2 is illustrated in Figure 9.5. By replacing a single processing element with an array of 8 by 8 PEs, a higher computation throughput can be achieved without increasing memory bandwidth. The function of the memory (i.e., the $\mathbf{L}$ matrix shift registers) in the diagram is to pulse data $l\_pipe_{jp}(j = 1, \ldots, n; p = 1, \ldots, m)$ through the array of cells. Then new data $l\_pipe_{ik}(i = 1, \ldots, n; k = 1, \ldots, m)$ are returned to memory in a rhythmic fashion. The crux of this approach is to ensure that once the data

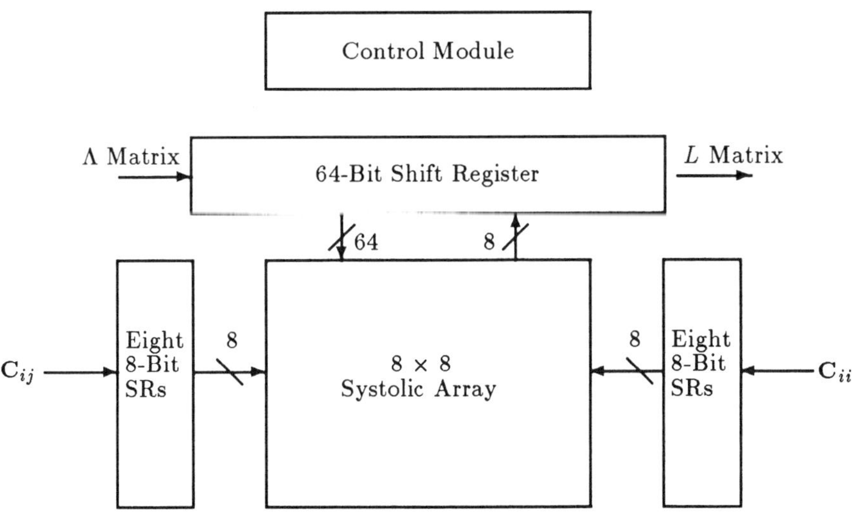


Figure 9.4: Circuit Block Diagram for the DRA2 System.

are brought out from the memory they can be used effectively at each cell they pass while being pumped through the entire array. To perform the parallel DRA2 computation, two cells [as illustrated in Figures 9.6 and 9.7] with almost identical logic and structure were used in constructing the entire array. The only difference is that the first cell performs the generation of the multiple broadcasting signals $b_k(k = 1, \ldots, m)$ for each row array while the second cell is only transparent to the passing of the $b_k$ signals. The construction of the multiprocessor SIMD array using these two cells is illustrated in Figure 9.8. According to Figure 9.3 and equations 9.10 to 9.13, these two cells are implemented in two levels of NOR gate combinational logic. Their PPL layouts can be easily identified in Figure 9.15. We see from Figures 9.6 and 9.7 that in essence the inner summation part of equation 4.4 is carried out in Cell-A and Cell-B, while the outer multiplication part of that equation is implemented in each horizontal array in Figure 9.8.

It is also noted from Figures 9.6 and 9.7 and 9.8 that the DRA2 architecture can readily be extended to different numbers of objects and labels. For example, a row array of the number $k$ in Figure 9.8 can be added in order to attach one more label for all objects, while a column array of the number $j$ in the same figure can be added to extend the system for one more object, provided that the corresponding changes inside each cell are made. The critical analysis indicates that the monolithic DRA2 systems for 8, 16, and 32 objects are technically implementable in the 3.0-$\mu$m NMOS, 2.0-$\mu$m and 1.2-$\mu$m CMOS, and 1.0-$\mu$m GaAs processes[22].

### Circuit Features and Design Techniques

In addition to designing the simple and regular cells, several efficient techniques, such as interleaved processing, multiple signal broadcasting, and self-timed synchronization, were applied to the implementation of the SIMD multiprocessor DRA2

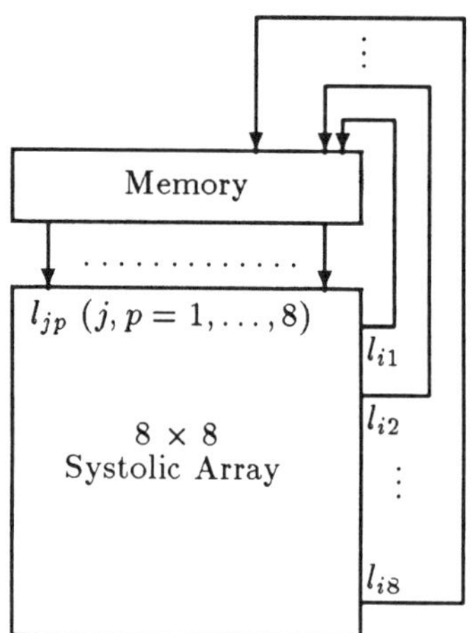


Figure 9.5: Basic principle of the parallel DRA2 system.

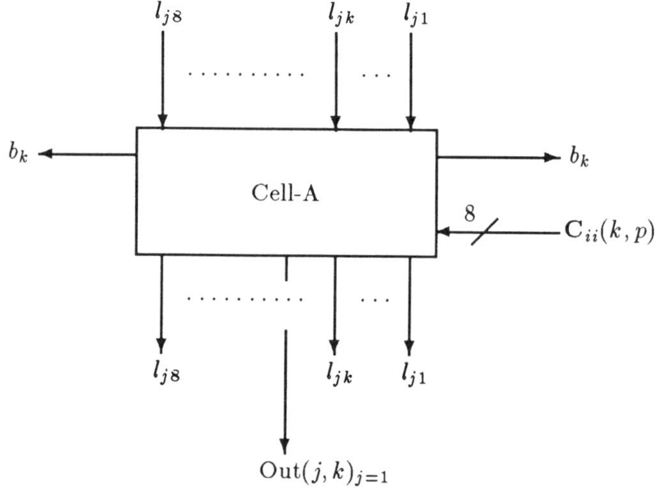


$$b_k = l_{jk}, \text{at column } j = 1 \tag{9.10}$$

$$\begin{aligned} \mathrm{Out}(j, k)_{j=1} &= \sum_{p=1}^{m} l_{jp} * \Lambda_{ij}(k, p) = \sum_{p=1}^{m} l_{jp} * l_{ik} * \mathbf{C}_{ii}(k, p) \\ &= \sum_{p=1}^{m} l_{jp} * b_k * \mathbf{C}_{ii}(k, p) = \sum_{p=1}^{m} (\overline{\overline{l_{jp}} + \overline{b_k} + \overline{\mathbf{C}_{ii}(k, p)}}) \end{aligned} \tag{9.11}$$

(a) Cell-A

Figure 9.6: A cells in multiprocessor DRA2 array.

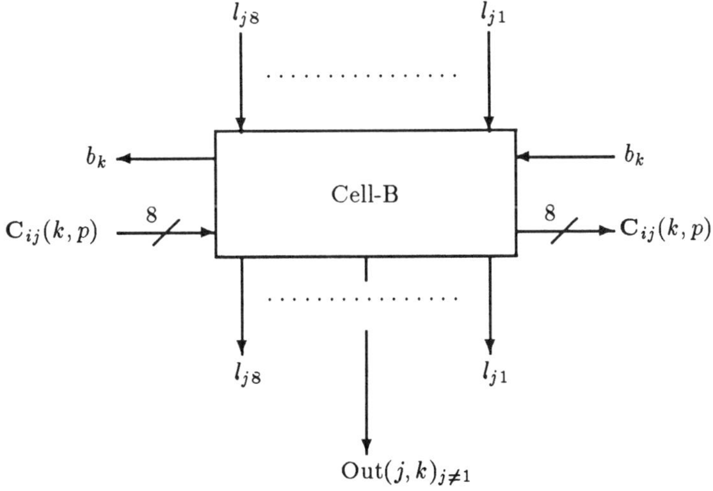


$$b_{k(\text{in})} = b_{k(\text{out})}, \text{ at columns } j \neq 1 \tag{9.12}$$

$$\begin{aligned}\text{Out}(j,k)_{j\neq 1} &= \sum_{p=1}^{m} l_{jp} * \Lambda_{ij}(k,p) = \sum_{p=1}^{m} l_{jp} * l_{ik} * \mathbf{C}_{ii}(k,p) \\ &= \sum_{p=1}^{m} l_{jp} * b_k * \mathbf{C}_{ii}(k,p) = \sum_{p=1}^{m} (\overline{\overline{l_{jp}} + \overline{b_k} + \overline{\mathbf{C}_{ii}(k,p)}})\end{aligned} \tag{9.13}$$

(b) Cell-B

Figure 9.7: B cells in multiprocessor DRA2 array.

architecture.

*Recursive Systolic Computation and Interleaved Processing*

Since the introduction of a time dimension in subsection 9.4.1, the multiprocessor SIMD array in Figure 9.8 possesses a time-varying characteristic which makes recursive systolic computation and interleaved processing possible. Let's focus on the first column ($j = 1$) array. It is clearly indicated that the first input vector, which is the $i$th row vector of $\mathbf{L}$, the labeling matrix, at the $(n-1)$th iteration, is fed into the first column of DRA2 array as:

$$(l_{j1}, l_{j2}, l_{j3}, \ldots, l_{jm})^t$$

where $j = 1, \ldots, n$. The corresponding output vector $\mathbf{L}_i$ of the multiprocessor SIMD array, which is the $i$th row vector of the $\mathbf{L}$ labeling matrix at the $n$th iteration, is generated:

$$(l_{i1}, l_{i2}, l_{i3}, \ldots, l_{im})^t$$

where $i$ is fixed at a time $t = i$. As time moves forward, the elements in the $L$ shift register have shifted from the left to the right in an 8-clock-pace fashion. For example, in the DRA2 system, at time $t = i = 1$, vector $\mathbf{L}_1$:

$$(l_{11}, l_{12}, l_{13}, l_{14}, l_{15}, l_{16}, l_{17}, l_{18})^t$$

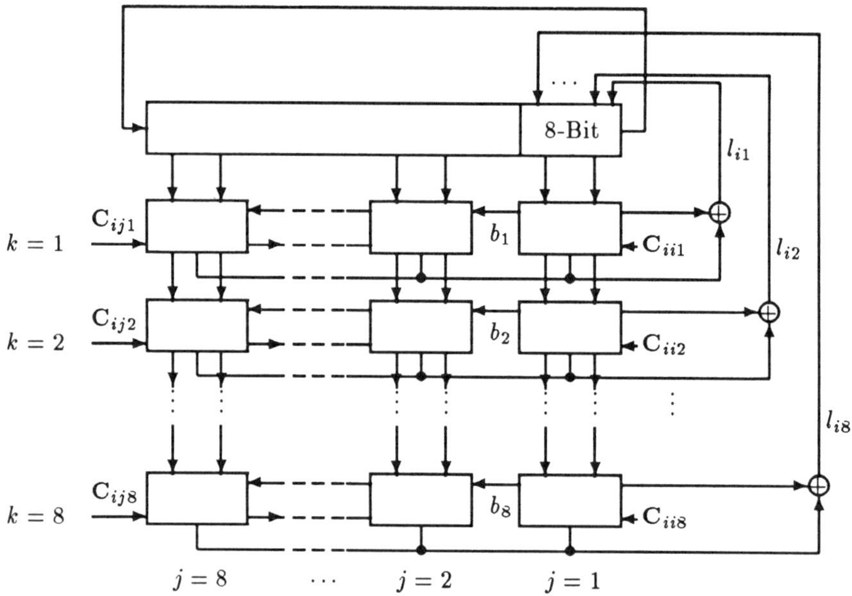


Figure 9.8: Construction of SIMD multiprocessor array using Cell-A and Cell-B.

is generated, and at time $t = i = 2$, vector $\mathbf{L}_2$:

$$(l_{21}, l_{22}, l_{23}, l_{24}, l_{25}, l_{26}, l_{27}, l_{28})^t$$

is generated, etc. as shown in Figure 9.9 for $l_{1i}$ and in Figure 9.10 for $l_{2i}$.

Each $\mathbf{L}_i$ vector is computed based on the interleaved utilization of the multiprocessor SIMD array, whereas eight $\mathbf{L}_i$ vectors form an entire computational wavefront of the $\mathbf{L}$ labeling matrix, of the $n$th relaxation iteration. Note that we use $n$ computing trees for generating $nm$ $l_{ik}$'s in $O(nm)$ time; we may also use $nm$ or $n$ computing trees to compute the same number of $l_{ik}$'s in $O(n)$ time.

*Multiple Signal Broadcasting*

The broadcasting technique is probably one of the most obvious ways to make multiple use of each input element. It plays an important role in making the parallel computation tree of Figure 9.3 implementable. Two multiple broadcasting schemes are used in the DRA2 architecture. In the first, $nm$ vertical broadcasting lines from each pipelining operand are connected to the bottommost leaves' node of each parallel computing tree (passing each $l$_pipe to $l$_root). Secondly, as depicted in Figures 9.6 and 9.7 and 9.11, Cell-A at column $j(=1)$ is used to jog signal $l_{ik}^{(n-1)}$ (which is the $n$th $l$_pipe$_{jp}$) and then propagate it horizontally from right to left through the entire row array. Thus, the output vector of the multiprocessor SIMD array, i.e., $(l_{i1}, l_{i2}, l_{i3}, l_{i4}, l_{i5}, l_{i6}, l_{i7}, l_{i8})$, can be generated simultaneously in a highly concurrent manner.

**Definition 9.4** *The second multiple data routing scheme for jogging $b_k$ at column $j = 1$, as illustrated in equations 9.8 and 9.10, and Figure 9.11, is defined as the* J-Pattern.

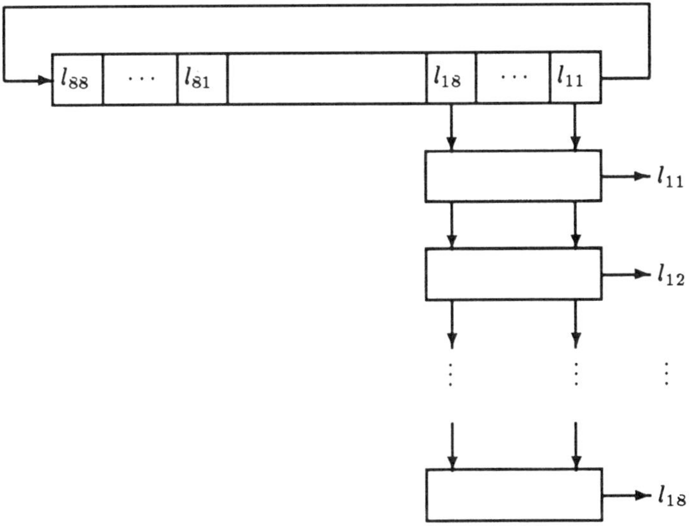


Figure 9.9: Computational wavefront pipelining at time 1.

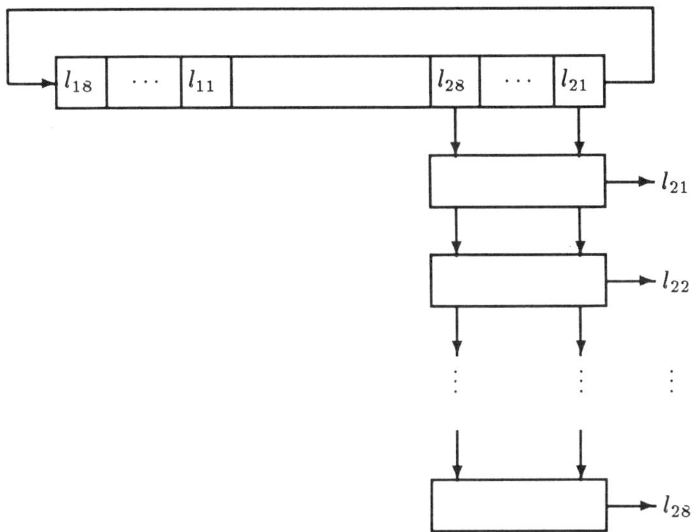


Figure 9.10: Computational wavefront pipelining at time 2.

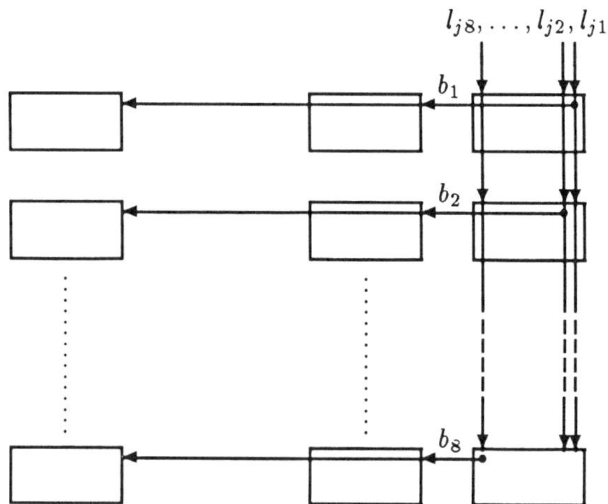


Figure 9.11: Broadcasting scheme for the $b_k$ signals.

*Cross Circle Associative Pipelining*

Much associativity is introduced through the associative circular pipelining in the DRA2 design. Refer to Figures 9.8 through 9.11, where two circular pipelining loops are implemented.

The first pipeline is designed for the $l_{jp}$ elements. This pipelining can be viewed in two parts, master pipelining part and slave pipelining part. The master pipelining refers to a serial horizontal circular pipelining of the $l_{jp}$ elements inside the main pipelining channel, LSR. While the master pipelining is going on, a parallel horizontal virtual circular pipelining is followed in $k$ row horizontal arrays. The parallel pipelining in DRA array is a slave pipelining of the master one which is supported by multiple signal broadcasting. Viewing Figure 9.12, the second pipelining for $l_{ik}$ is a parallel vertical circular pipelining which crosses the serial horizontal pipelining at the rightmost first $m$-bit of the main pipelining channel, to update the old labels in the channel with new labels.

In addition to providing massive parallelism, an extra advantage of these pipelining schemes is that it decreases the routing and control complexities.

*Self-Timed Synchronization and Tagged-Bit Control*

By using recursive systolic computation and interleaved processing, the computational task has been decomposed into the smallest computing piece, $\mathbf{L}_i$. To compute each vector $\mathbf{L}_i$, the globally synchronized multiprocessor SIMD array of Figure 9.8 is used. For completing the entire relaxation computation, this synchronous array is imbedded into a self-timed system. The self-timed asynchronous control may be costly in terms of extra hardware and delay in each element, but it has the advantage that the time required for a communication event between two elements is independent of the size of the entire system[56]. Also, it is easy to design and validate a self-timed state machine in PPL methodology[33].

Among the 64-bit $\mathbf{L}$ shift registers, the rightmost first 8-bit SR is one which is able to parallel load the $n$th output vector from the multiprocessor SIMD array in order to update the current $(n-1)$th $\mathbf{L}_i$ row vector. This iteration and updating

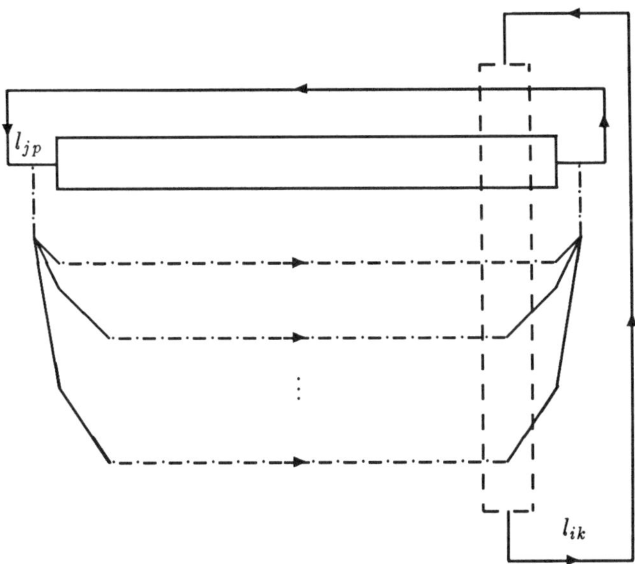


Figure 9.12: Associative circular pipelining.

process is the core of the relaxation process. To sense the completion of computation, a Comparator is built on top of the first 8-bit SR. If two vectors are equal, a *row-eq* signal of 1 is produced and stored into *8-bit States SR* of the control module; otherwise a 0 signal is sent. As soon as the state register gets eight 1's, which means equality is reached, an *all-eq* signal is issued to the FSM. Since the control processes in this system are based on the data validity of a *control data flow*, reliable and fast execution in a data-driven environment is created. The control mechanism used in the parallel DRA2 architecture is shown in Figure 9.13.

To ensure that the iteration cycle completes at the end of $nm$ cycles, a *tagged-bit* is derived from an ANDed term of both the $l_{11}$ bit and the 64th-count of the timer, which has served as a reliable alignment signal for computation in the control flow. In Figure 9.14 the state graph of the FSM is illustrated.

### 9.4.3 Performance Evaluation

With the parallel tree-structured reformulation and multi-dimensional tree-root pipelining in its system design, the DRA2 architecture takes advantage of a high degree of pipelining and multiprocessing. *It eliminates the need for storing and computing each element associated with the object pairs' matrixes* $\Lambda_{ij}(k,p)$. One-half of the $O(n^2m^2)$ space is eliminated; the other half is reserved for input data $\mathbf{C}_{ij}(k,p)$, the label pairs' compatibility matrixes. In certain DRA computations, as shown in the *region coloring problem*, the space complexity can even be reduced to $O(nm)$; i.e., only an $nm$-bit shift register is required to store all $nm$ intermediate (or original) labeling elements. Obviously, the DRA2 computation during each iteration only takes $O(nm)$ clock cycles. Assuming a clock cycle is about 120 ns (using a 3-$\mu$m NMOS process and the PPL design methodology). DRA2 performs the DRA computation in microseconds, and in the worst case, i.e., multiplying the maximum possible iterations, $O(nm)$, in milliseconds.

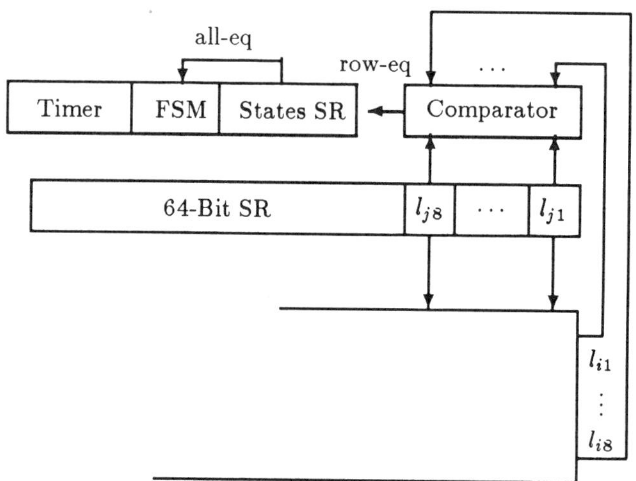


Figure 9.13: Self-timed synchronization.

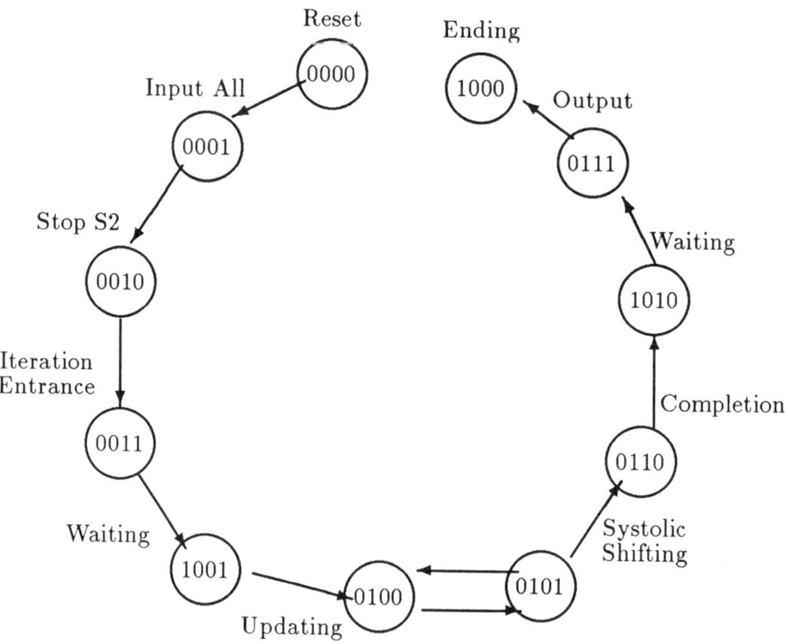


Figure 9.14: State graph of the finite state machine.

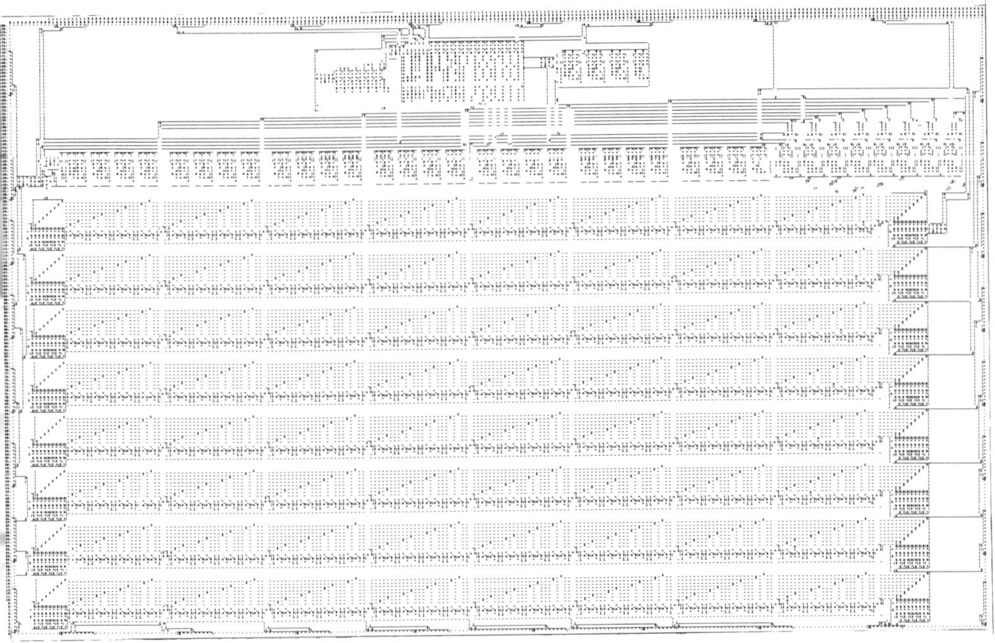

Figure 9.15: The PPL layout of the multiprocessor SIMD DRA2 system.

### 9.4.4 The PPL VLSI Layout for the DRA2 System

The DRA2 chip was built by assembling the four functional blocks in Figure 9.4 using PPL (Path Programmable Logic) tools at the University of Utah. Since parallel computation and the multiprocessor SIMD array greatly simplify the design difficulties, the PPL layout is very simple and straightforward. An overview of the complete PPL layout, which is a PPL mapping of the block *floor plan* in Figure 9.4, is shown in Figure 9.15; a photomicrograph of the DRA2 chip is shown in Figure 9.16.

For details associated with the complete system design, the simulation results, interfacing strategy with host computer, timing and wiring delay analysis, testing, pinout description, and fabrication of the DRA2 chip, see References 60, 91 and 92.

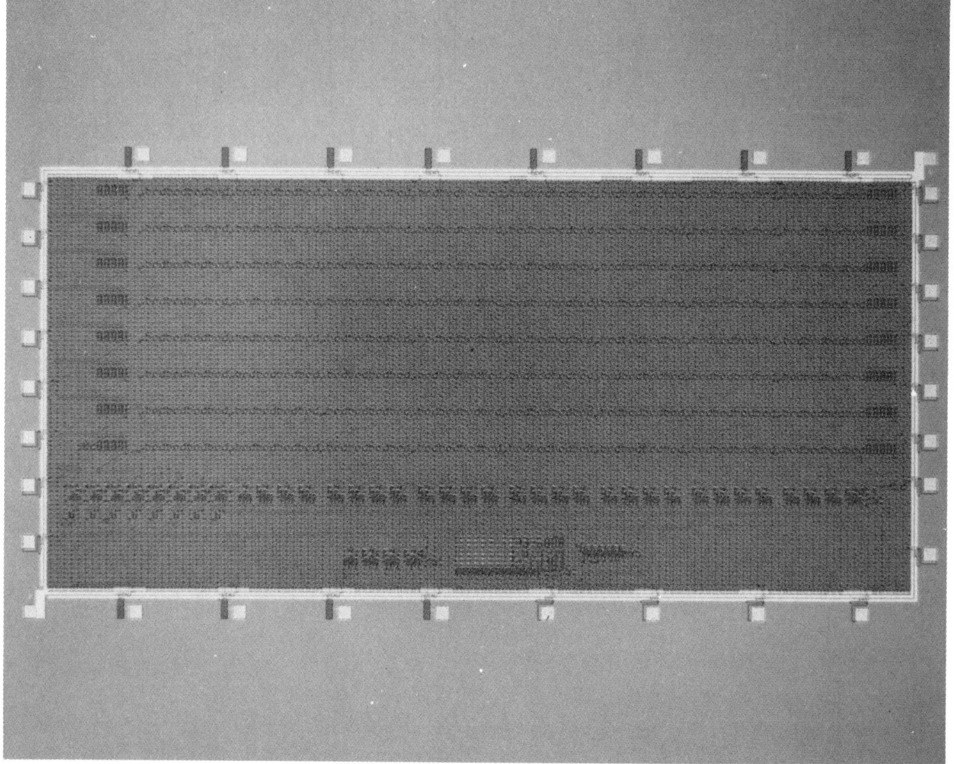

Figure 9.16: The photomicrograph of the DRA2 chip.

## 9.5 The Dynamic Architectural Wavefront

Compared to the DRA1 design, the DRA2 system achieves a very impressive performance improvement in terms of speed, size, and memory access requirements. However, there are still several bottlenecks in developing a fast general-purpose DRA architecture.

### 9.5.1 Complexity Issues Revisited

- *Time complexity*: The $O(nm)$ time of the DRA2 system is suitable only for a small number of objects, though the number of objects and labels in the DRA2 architecture can be extended to a larger number of objects and labels. The larger the number of objects and labels are, the slower the DRA2 system becomes. It is desirable that a much faster DRA architecture be designed for real-time processing of a larger number of objects and labels.

- *Data preprocessing complexity*: One of the remarkable characteristics of the DRA problem is that a large number of data, such as $O(n^2m^2)$ elements of label pairs' compatibility matrixes $\mathbf{C}_{ij}(k,p)$, must be used during each interphase DRA iteration. In addition to this, the data ordering and format which might be processed during the load-in time must be arranged to support efficient DRA computation. Following assumption 3, the term *data preprocessing complexity (DPC)* is used here to refer to the complexity issues which arise in the analysis of the time and space, as well as hardware complexities in this kind of data-preprocessing-intensive VLSI computation. Some labeling solutions avoid this problem by assuming that the data is always ready for computation.

- *Data pattern routing complexity*: One of the proposed DRA3 architectures is shown in Figure 9.20, in which the data routing complexity in the horizontal direction between two DRA modules is of $O(m^2)$; a total of $O(nm^3)$ connections are required for routing $nm$ DRA modules. The data routing complexity becomes another dominant factor in designing the DRA3 system as the number of objects and labels increases.
- *Fabrication difficulty*: Although we have separated the main pipelining channel and have used the larger buffers to drive between shift registers, its characteristic gets worse as the numbers of $n$ and $m$ increase.

### 9.5.2 $C_{ij}(k,p)$-Pattern Analysis

The DRA3 architecture overcomes these difficulties based on the use of the dynamically configured architectural wavefront on the DRA3 switch lattice. It runs the DRA computation in $O(n)$ time without the extra requirement for data preprocessing hardware and without requiring the horizontal routing among the DRA modules, provided that only an $O(nm^2)$ number of switches must be added to the original multiple broadcasting wires of the DRA2 system.

In fact, computing time, data preprocessing complexity, and data routing complexity form a coherent mixture in the DRA3 system design. We illustrate its major architectural concepts beginning with the analysis of $\mathbf{C}_{ij}(k,p)$ distribution patterns in the case of $n = m = 3$.

Referring to equation 4.4 and Figure 9.9, the first computational wavefront for computing the $n$th labels $l_{ik}^{(n)}$ is formed at time $t = i = 1$ for $i = 1$ and $k = 1,2,3$ as:

$$
\begin{aligned}
l_{1k}^{(n)} \leq l_{1k}^{(n-1)} &* \left[l_{11}^{(n-1)} * \Lambda_{11}(k,1) + l_{12}^{(n-1)} * \Lambda_{11}(k,2) + l_{13}^{(n-1)} * \Lambda_{11}(k,3)\right] \\
&* \left[l_{21}^{(n-1)} * \Lambda_{12}(k,1) + l_{22}^{(n-1)} * \Lambda_{12}(k,2) + l_{23}^{(n-1)} * \Lambda_{12}(k,3)\right] \\
&* \left[l_{31}^{(n-1)} * \Lambda_{13}(k,1) + l_{32}^{(n-1)} * \Lambda_{13}(k,2) + l_{33}^{(n-1)} * \Lambda_{13}(k,3)\right]
\end{aligned}
\tag{9.14}
$$

A snapshot of the $\mathbf{C}_{1j}(k,p)(j,k,p = 1,2,3)$ matrixes data distribution pattern which is matched with the $l_{1k}^{(n)}$ wavefront at $t = 1$ is given in Figure 9.17. At time $t = 2$, a computational wavefront for generating $l_{2k}^{(n)}$ is produced. For $i = 2$ and $k = 1,2,3$ as:

$$
\begin{aligned}
l_{2k}^{(n)} \leq l_{2k}^{(n-1)} &* \left[l_{11}^{(n-1)} * \Lambda_{21}(k,1) + l_{12}^{(n-1)} * \Lambda_{21}(k,2) + l_{13}^{(n-1)} * \Lambda_{21}(k,3)\right] \\
&* \left[l_{21}^{(n-1)} * \Lambda_{22}(k,1) + l_{22}^{(n-1)} * \Lambda_{22}(k,2) + l_{23}^{(n-1)} * \Lambda_{22}(k,3)\right] \\
&* \left[l_{31}^{(n-1)} * \Lambda_{33}(k,1) + l_{32}^{(n-1)} * \Lambda_{33}(k,2) + l_{33}^{(n-1)} * \Lambda_{33}(k,3)\right]
\end{aligned}
\tag{9.15}
$$

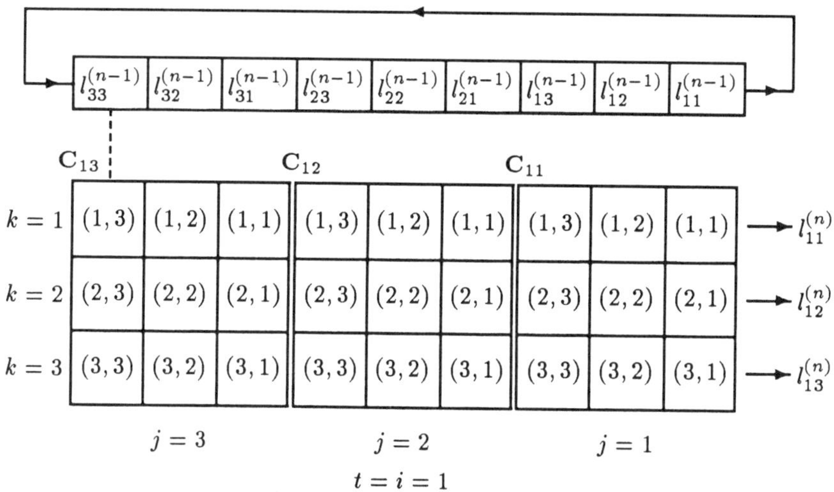


Figure 9.17: The $\mathbf{C}_{1j}(k,p)$ distribution pattern for generating $l_{1k}$'s.

The $\mathbf{C}_{2j}(k,p)(j,k,p = 1,2,3)$ matrixes data distribution pattern at this time is shown in Figure 9.18. From Figures 9.17 and 9.18, several criteria for DRA3 architecture are derived.

1. At time $t = i$, to compute each new labeling vector $\mathbf{L}_i$, different computational wavefronts are formed and the associated $\mathbf{C}_{ij}(k,p)$ matrixes are distributed. More accurately, at time $t$, each matrix $\mathbf{C}_{tj}(k,p)$ elements (where $t,j = 1,\ldots,n$ and $k,p = 1,\ldots,m$) stored in the $k$th row of a parallel memory in Figures 9.17 and 9.18 are retrieved. Thus a parallel accessing (parallel row-readable) RAM is required.

2. It is clearly indicated that the $\mathbf{C}_{ij}(k,p)$ elements associated with each dynamically pipelined operand in the main tree-root pipelining channel are one unit of $j$ skewed and horizontally circulated. The cyclic shifting of the bits of $\mathbf{C}_{ij}(k,p)$-patterns' elements to the right for $m$-bit units corresponds to a *block-data shuffling* operation. We summarize and extend all distributed $\mathbf{C}$ matrixes' patterns in Figure 9.19. (Each symbol actually is a matrix distributed in Figures 9.17 and Figures 9.18.)

The first requirement requires a simple associative processing, i.e., parallel accessing RAMs; while the second criterion demands higher-order design complexity in time and data preprocessing, as well as in data routing.

A candidate design for the DRA3 system, where $n = m = 8$, is presented in Figure 9.20. The DRA3 system is made up of using 8 by 8 DRA modules. Each module consists of a PE cell (same structure as Cell-A and Cell-B in DRA2 system) and a local parallel accessing RAM for the $\mathbf{C}_{ij}(k,p)$-pattern.

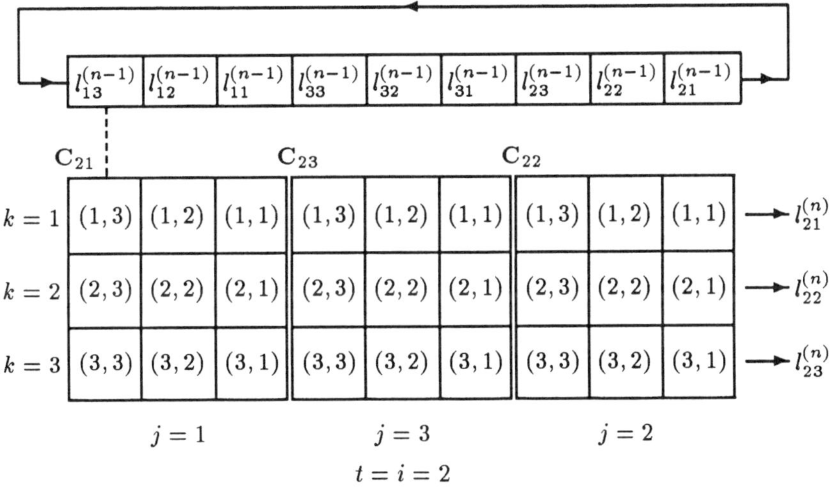


Figure 9.18: The $\mathbf{C}_{2j}(k,p)$ distribution pattern for generating $l_{2k}$'s.

$$\begin{array}{llll} t=1: & \mathbf{C}_{13}(k,p) & \mathbf{C}_{12}(k,p) & \mathbf{C}_{11}(k,p) \\ t=2: & \mathbf{C}_{21}(k,p) & \mathbf{C}_{23}(k,p) & \mathbf{C}_{22}(k,p) \\ t=3: & \mathbf{C}_{32}(k,p) & \mathbf{C}_{31}(k,p) & \mathbf{C}_{33}(k,p) \end{array}$$

Figure 9.19: A circularly skewed $\mathbf{C}_{ij}(k,p)$-pattern when generating $l_{ik}$'s.

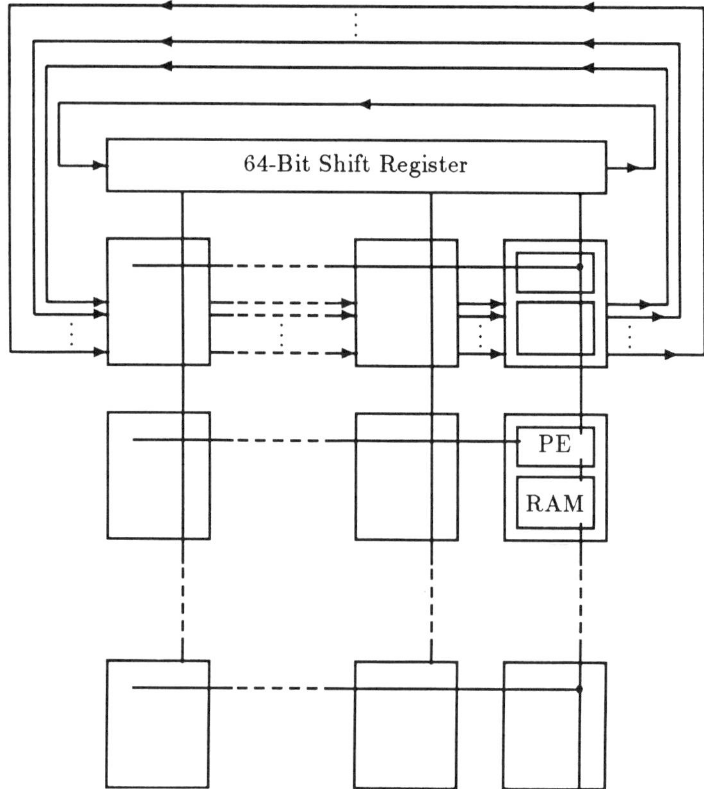


Figure 9.20: A candidate design for DRA3 system.

A preshuffling chip is implemented and fabricated using a 3-$\mu$m NMOS PPL design methodology in order to make the $\mathbf{C}_{ij}(k,p)$-pattern the same format as in Figure 9.19[5]. However, the preprocessing hardware costs $O(b^3)$ time (noting that $b$ is the number of bits to be shuffled) and takes more than half of the chip size, while a difficult interchip routing problem remains unsolved.

### 9.5.3 A Dynamically Configurable DRA3 Architecture

The configurable, highly parallel computer system is a multiprocessor architecture that provides a programmable interconnection structure integrated with the PEs[77]. The idea is that the computer processing begins with the controller broadcasting a command to all switches to invoke a particular *static* configuration setting. The design of the DRA3 system has revealed that by adopting an advanced *dynamic* configuring strategy on the DRA3 switch lattice, not only can the DRA computational wavefront be generated while retaining the benefits of uniformity and locality that DRA-PE exploits, but also the combined limitations imposed in upgrading DRA3 architecture can be completely eliminated.

### The Architectural-Computational Wavefront (ACW) Notation

We rewrite Weiser and Davis's definition of wavefront[91]:

**Definition 9.5** *A* wavefront, *denoted as A, represents an ordered set of data elements:*

$$\{a(1,m), a(2,m), \ldots, a(N-1,m), a(N,m)\}$$

*where $m$ is the "time" subscript. The elements $a(I,m)$ for all $m$ belong to the Ith data stream. For simplicity, the "time" subscript in the elements of a wavefront is omitted and $a(i,m)$ will be simply represented as $a(i)$.*

We extend the definition as:

**Definition 9.6** *The* architectural-computational wavefront (ACW) notation*[27] consists of two different wavefronts. The* computational wavefront *is an ordered set of data elements: $\{c(1,i+t), c(2,i+t), \ldots, c(N,i+t)\}$, where $i+t$ is the spatial-temporal index in equation 9.8. The* architectural wavefront *is an ordered set of architectural configurations: $\{a(1,j+t), a(2,j+t), \ldots, a(N,j+t)\}$, where $j+t$ is the spatial-temporal index in equation 9.9.*

The following distinctions are made which are necessary for building a DRA3 system:

1. The computational wavefront dynamically progresses on a static architecture, as the $ST$ indexes increase. The architectural wavefront progresses, by dynamically configuring the system architecture, in a way against the computational wavefront, as the $ST$ indexes increase. For instance, as we have seen that the $ST$ index in DRA2 forms a dynamic DRA computational wavefront on a static DRA2 architecture, we see below that the $ST$ index of DRA3 generates a dynamic DRA3 architectural wavefront for a virtual static DRA computation.
2. The equivalence between computational wavefront and architectural wavefront holds. Thus either one characterizes both of them.
3. Both architectural and computational wavefront notation have spatial parameters, such as $i$ and $j$, and temporal parameters, such as $t$, so that both can be manipulated on the space and/or time domains.
4. The conventional wavefront notation assumes a uniform progression of the data stream. This restriction is eliminated by combining $ST$ indexes with ACW notation, as it is obvious that these indexes may behave synchronously or asynchronously.
5. The ACW notation first differentiates the computational wavefront and architectural wavefront and is more suitable for exploring architectural configurability and the synthesis of highly configurable system.

Referring to equations 4.4, 9.8 and 9.9, there is always a data dependence relation among these arguments with respect to time $t$ and positions $k$ and $j$, that:

$$l\_\text{pipe}_{i+t,k}^{(n)} \leftarrow \sum_{p=1}^{m} l\_\text{root}_{i+t,k}^{(n-1)} l\_\text{root}_{j+t,p}^{(n-1)} \mathbf{C}_{i+t,j+t}(k,p) \qquad (9.16)$$

In Figure 9.8 we see that this data dependency equation is topologically mapped onto silicon in terms of the parallel tree structure. The vertical broadcasting of the $l$_pipe operands from pipelined channel to DRA array is described by a routing function:

$$l\_\mathrm{pipe}_{i+t,k}^{(n-1)} \quad \leftarrow \quad l\_\mathrm{root}_{j+t,p}^{(n-1)} \updownarrow \mathbf{C}_{i+t,j+t}(k,p) \tag{9.17}$$

where the symbol $\updownarrow$ denotes an alignment relation such that $A \updownarrow B$ if and only if $A$ and $B$ have the same physical column position of $j$. The *J-pattern* of definition 4 is routed by equation:

$$l\_\mathrm{root}_{j+t,p}^{(n-1)} \quad \rightarrow \quad l\_\mathrm{root}_{i+t,k}^{(n-1)} \updownarrow C_{i+t,j+t}(k,p) \tag{9.18}$$

The DRA computation is valid if and only if these two dependency equations are true at any time $t$ and positions $k$ and $j$, as well as the iteration times $n$. Equation 9.17 stands for a static network in the DRA3 architecture. Equation 9.18 is a virtually static network; a potential speed-up in the DRA3 computation can be reached by dynamic configuring equation 9.17 in an ACW notation.

**Dynamically Configuring the DRA3 Architectural Wavefront**

Though symbol $j$ represents the horizontal spatial shifting (in Figures 9.17 and 9.18 while the symbol $t$ represents the temporal unit, the computational wavefront progression of $l\_\mathrm{root}_{j+t,p}$ proceeds [referring to equations 9.8 and 9.9] as the increasing of $j+t$, where the combined parameter $j+t$ must not be associated with any dimensions. By definition 9, it is obviously correct that the $l_{ik}$ computational wavefront moves from left to right, as the $ST$ indexes increase and is equivalent to that of the *J-pattern* (one configuration of the architectural wavefront) shifts in the reverse horizontal direction, as the $ST$ indexes increase. The DRA computation can be performed by manipulating *either* a computational wavefront *or* configuring dynamically an architectural wavefront.

In Figure 9.21, we create a dynamically configurable DRA3 architectural wavefront by building an algorithm-configured switch lattice. A total number of $O(nm^2+nm)$ switches are added between each vertical wire and horizontal wire. In Figure 9.21, there are $nm$ identical DRA modules denoted by $M_{jk}(j=1,\ldots,n;k=1,\ldots,m)$. Let $SN_{jk}$ denote a switching node which connects vertical broadcasting wire $l_{jk}$ and horizontal broadcasting wire $b_k$ inside of each module. Let $BS_{jk}$ denote a bus switch which connects the labeling output of the multiprocessor SIMD array to the operands of the "pipelining channel," noting that the pipelining channel is now static and is replaced with an $O(nm)$-bit RAM. These switches are programmed by a *self-timed system controller*, as is shown in the DRA2 circuit, to avoid the synchronization problem caused by time delay during the broadcasting on the switch lattice. For DRA3 computation, we give:

**Definition 9.7** *We define the* DRA3 architectural wavefront *to be the interconnection patterns of the DRA3 switch lattice, of which both the switches $SN_{jk}$ for a column $j$, i.e., the J-pattern at a specified column $j$, and switches $BS_{jk}$ for that column $j$ are invoked simultaneously, where index $j$ is defined in equation 9.9.*

During the DRA3 computation, at index $j(\mathrm{mod}\ n)$, the DRA3 architectural wavefront shifts from right to left in $O(n)$ clock cycles, therefore, an entire DRA computation, which is virtually static, is dynamically generated.

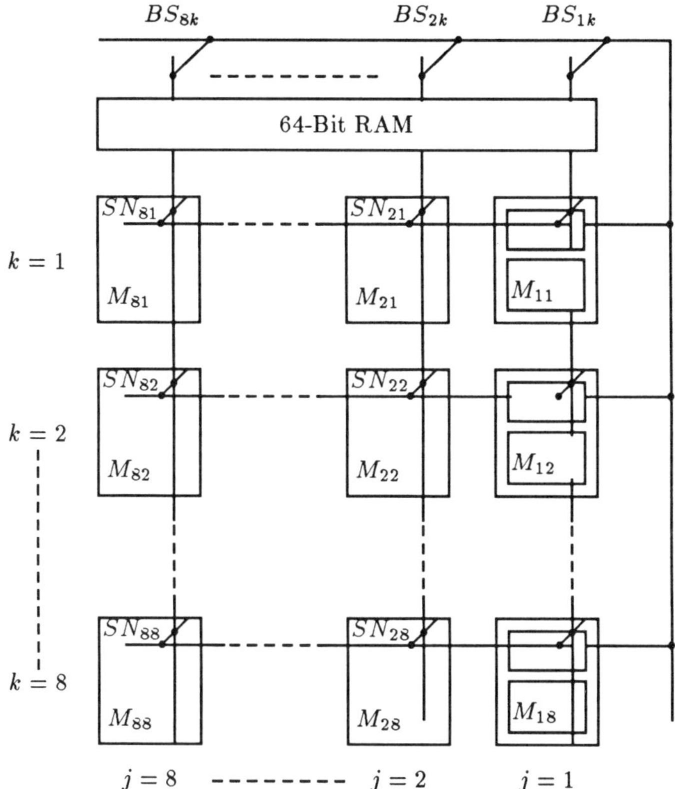


Figure 9.21: The DRA3 architecture.

It would now be of interest to examine whether the complexity issues in Section 9.5.1 are solved upon building the DRA3 architecture. First, a time upper bound of $O(n)$ is reached. Secondly, there is no particular hardware support and no special requirements for data-preprocessing. Under dynamic parallel configuring of DRA3, the $\mathbf{C}_{ij}(k,p)$-pattern is naturally distributed in natural matrix index order in the original parallel memory. For an intuitive understanding, we redraw the $\mathbf{C}_{ij}(k,p)$-pattern of Figure 9.19 in Figure 9.22 which is associated with DRA3 computation strategy. The third problem, an intensive multiple data pattern routing requirement is eliminated by adding $O(nm^2 + nm)$ switches on the DRA3 switch lattice. As for the long pipelining channel, it is replaced with $O(nm)$ bits of smaller and more

| | | | |
|---|---|---|---|
| $t = 1$: | $\mathbf{C}_{13}(k,p)$ | $\mathbf{C}_{12}(k,p)$ | $\mathbf{C}_{11}(k,p)$ |
| $t = 2$: | $\mathbf{C}_{23}(k,p)$ | $\mathbf{C}_{22}(k,p)$ | $\mathbf{C}_{21}(k,p)$ |
| $t = 3$: | $\mathbf{C}_{33}(k,p)$ | $\mathbf{C}_{32}(k,p)$ | $\mathbf{C}_{31}(k,p)$ |

Figure 9.22: A $\mathbf{C}_{ij}(k,p)$-pattern under dynamic reconfiguration of DRA3.

reliable RAM cells, under the dynamic data routing pattern. Finally, since $\mathbf{C}_{ij}(k,p)$-pattern is designed for $O(n^2)$ arbitrary $m$ by $m$ $\mathbf{C}$ matrixes, the DRA3 system is good for any general-purpose DRA computation.

## 9.6 Conclusions

VLSI architectures for speeding up the computation of the discrete relaxation algorithm are described. The key issues are a multi-dimensional tree-root pipelining scheme and a technique to generate the DRA data pattern and DRA architectural wavefront simultaneously. DRA2 and DRA3 multiprocessor architectures both achieve high performance by dynamically matching the internal structure of the parallel computing machine to the implied structure of the data pattern which is to be processed. The implementations of these architectures offer much greater processing performance than general-purpose processors.

Currently several optimal, parallel DRA algorithms, M-DRA3 and DRA5, are being developed. Three new computer architectures were designed,implemented, and some of them fabricated. These algorithms and architectures are *optimal* in the sense that both the optimal algorithm lower bound, $O(1)$, and a very small time constant are reached[26]. Further research in this area is to imbed a higher degree of flexibility in architecture design by allowing programmability in cells as well as reconfigurability of cell interconnections, and to imbed a larger amount of massive parallelism in DRA data pattern generation and computation, and to build a wafer scale computer system for real-time processing of various large size DRA problems.

## 9.7 Acknowledgment

Jun Gu would like to acknowledge support for this research in part by an ACM-IEEE Design Automation Scholarship Award, an ACM Special Academic Award, and a University of Utah Research Fellowship.

# Bibliography

[1] D.N. de G. Allen. *Relaxation Methods.* McGraw-Hill, New York, 1954.

[2] H. Barrow and J. Tennenbaum. *MSYS: A System for Reasoning about Scenes.* Technical Report AI Tech. Report 121, SRI, 1976.

[3] D.A. Bell. Decision Trees, Tables, and Lattices. In B.G. Batchelor, editor, *Pattern Recognition Ideas in Practice*, pages 119–141, Plenum Press, New York, 1978.

[4] A. Borning. The Programming Language Aspects of ThingLab, a Constraint-Oriented Simulation Laboratory. *ACM Transactions on Programming Languages and Systems*, 3(4):353–387, October 1981.

[5] W. Burger. A Virtual Shuffle Dynamic RAM Controller. In K. Smith, editor, *Fourth Annual University of Utah VLSI Contest*, University of Utah, April, 1986.

[6] M. Clowes. On Seeing Things. *Artificial Intelligence*, 2:79–116, 1971.

[7] Martin C. Cooper. An Optimal k-Consistency Algorithm. September, 1988. unpublished manuscript.

[8] L. Davis and Thomas C. Henderson. Hierarchical Constraint Processes for Shape Analysis. *IEEE Transactions on Pattern Analysis and Machine Intelligence*, PAMI-3(3):265–277, May 1981.

[9] L. Davis and A. Rosenfeld. Hierarchical Relaxation for Waveform Parsing. In *Computer Vision Systems*, pages 101–109, Academic Press, New York, 1978.

[10] R. Dechter and J. Pearl. The Anatomy of Easy Problems: A Constraint Satisfaction Formulation. In *IJCAI-85*, pages 1066–1072, Los Angeles, California, 1985.

[11] R. Dechter and J. Pearl. Network-Based Heuristics for Constraint-Satisfaction Problems. *Artificial Intelligence*, 34:1–38, 1988.

[12] E. W. Dijkstra. Co-operating Sequential Processes. In F. Genuys, editor, *Programming Languages, NATO Advanced Study Institute*, chapter 2, pages 43–112, Academic Press, New York, 1968.

[13] Olivier Faugeras and Marc Berthod. Scene Labeling: An Optimization Approach. *Pattern Recognition*, 12:339–347, 1980.

[14] J. Feder. Plex Languages. *Information Sciences*, 3:225–241, 1971.

[15] H. Freeman and L. Davis. A Corner-Finding Algorithm for Chain Coded Curves. *IEEE Transactions on Computers*, C-26:297–303, March 1977.

[16] E. Freuder and M. Quinn. Taking Advantage of Stable Sets of Variables in Constraint Satisfaction Problems. In *IJCAI-85*, pages 1076–1078, Los Angeles, California, 1985.

[17] E.C. Freuder. *Synthesizing Constraint Expressions*. Technical Report AI Memo 310, MIT, July 1976.

[18] K.S. Fu. *Syntactic Methods in Pattern Recognition*. Volume 112 of *Mathematics in Science and Engineering*, Academic Press, New York, 1974.

[19] K.S. Fu and B.K. Bhargava. Tree Systems for Syntactic Pattern Recognition. *IEEE Transactions on Computers*, C-22:1087–1099, December 1973.

[20] V. Gallo. A Program for Grammatical Pattern Recognition Based on the Linguistic Method of the Description and Analysis of Geometrical Structures. In *IJCAI-75*, pages 628–634, Tbilisi, Georgia, USSR, 1975.

[21] John Gaschnig. *Performance Measurement and Analysis of Certain Search Algorithms*. Technical Report CMU-CS-79-124, Carnegie-Mellon University, Pittsburgh, Pennsylvania, May 1979.

[22] M. Gaspar. Discussion of the PPL Design Methodology and MOSIS Fabrication Technology. 1986. Private Communication.

[23] Rafael C. Gonzalez and Michael G. Thomason. *Syntactic Pattern Recognition*. Volume 14 of *Applied Mathematics and Computation*, Addison-Wesley, Reading, Massachusetts, 1978.

[24] J. Gosling. *Algebraic Constraints*. PhD thesis, Carnegie-Mellon Univserity, Pittsburgh, Pennsylvania, May 1983.

[25] D. Gries. *Compiler Construction for Digital Computers*. John Wiley, New York, 1969.

[26] J. Gu. Parallel Consistent Labeling Computer Architectures. March, 1988. Ph.D Thesis Proposal.

[27] J. Gu, W. Wang, and Thomas C. Henderson. A Parallel Architecture for Discrete Relaxation Algorithm. *IEEE Transactions on Pattern Analysis and Machine Intelligence*, PAMI-9(6):816–831, 1987.

[28] A. Guzman. *Computer Recognition of Three-Dimensional Objects in a Visual Scene*. PhD thesis, MIT, Cambridge, Massachusetts, 1968.

[29] C.C. Han and C.H. Lee. Comments on Mohr and Henderson's Path Consistency Algorithm. *Artificial Intelligence*, 36:125–130, 1988.

[30] R. Haralick, L. Davis, A. Rosenfeld, and D. Milgram. Reduction Operations for Constraint Satisfaction. *Information Sciences*, 14:199–219, 1978.

[31] R. Haralick and L. Shapiro. The Consistent Labeling Problem I. *IEEE Transactions on Pattern Analysis and Machine Intelligence*, PAMI-1(2):173–183, April 1979.

[32] R.M. Haralick and G. Elliot. Increasing Tree Search Efficiency for Constraint Satisfaction Problems. *Artificial Intelligence*, 14:263–313, 1980.

[33] A. B. Hayes. Self-Timed IC Design with PPL's. In R. E. Bryant, editor, *Proceedings of the Third CalTech Conference on VLSI*, pages 257–273, Computer Science Press, Rockville, Maryland, 1983.

[34] Thomas C. Henderson. *Hierarchical Constraint Processes for Shape Analysis*. PhD thesis, The University of Texas, Austin, Texas, December 1979.

[35] Thomas C. Henderson. A Note on Discrete Relaxation. *Computer Vision, Graphics, and Image Processing*, 28:384–388, 1984.

[36] Thomas C. Henderson and L. Davis. Hierarchical Models and Analysis of Shape. *Pattern Recognition*, 14(1-6):197–206, May 1981.

[37] Thomas C. Henderson and Ashok Samal. Parallel Split-Level Relaxation. *International Journal of Pattern Recognition and Artificial Intelligence*, to appear, 1988.

[38] W. Daniel Hillis. *The Connection Machine*. The MIT Press, Cambridge, Massachusetts, 1985.

[39] G. Hinton. *Relaxation and Its Role in Vision*. PhD thesis, University of Edinbourgh, Edinbourgh, December 1977.

[40] B.K.P. Horn and B.G. Schunck. Determining Optical Flow. *Artificial Intelligence*, 17:185–203, 1981.

[41] S. Horowitz. Peak Recognition in Waveforms. In K.S. Fu, editor, *Syntactic Pattern Recognition Applications*, pages 1–31, Springer-Verlag, Berlin, 1977.

[42] D.A. Huffman. Impossible Objects as Nonsense Sentences. In B. Meltzer and D. Michie, editors, *Machine Intelligence*, pages 295–333, Edinburgh Press, Edinburgh, 1971.

[43] R.A. Hummel and S.W. Zucker. On the Foundations of Relaxation Labeling Processes. *IEEE Transactions on Pattern Analysis and Machine Intelligence*, PAMI-5(3):267–286, May 1983.

[44] BBN Laboratories Inc. *Butterfly Parallel Processor Overview*. Report number 6148, March 6 1986.

[45] D. Knuth. Semantics of Context-Free Grammars. *Mathematical Systems Theory*, 2(2):127–145, 1968.

[46] William A. Kornfeld. *The Use of Parallelism to Implement a Heuristic Search*. Technical Report A.I. Memo No. 627, MIT, AI Lab, March 1981.

[47] William A. Kornfeld. *Using Parallel Processing for Problem Solving*. Technical Report A.I. Memo No. 561, MIT, AI Lab, December 1979.

[48] D. Ku. DRA1 Chip Implementation Report, Class Project, University of Utah. VLSI Project.

[49] K. Kulkarni and L. Kanal. An Optimization Approach to Hierarchical Classifier Design. In *Proc. of the International Joint Conference on Pattern Recognition*, pages 459–466, Los Angeles, California, 1976.

[50] H. T. Kung. Putting Inner Loops Automatically in Silicon. In Tosiyasu L. Kunii, editor, *Lecture Notes in Computer Science* Vol. 163*: VLSI Engineering*, pages 70–104, Springer-Verlag, Berlin, 1985.

[51] J.-L. Lauriere. A Language and a Program for Stating and Solving Combinatorial Problems. *Artificial Intelligence*, 10:29–127, 1978.

[52] A.K. Mackworth. Consistency in Networks of Relations. *Artificial Intelligence*, 8:99–118, 1977.

[53] A.K. Mackworth and E.C. Freuder. The Complexity of Some Polynomial Network Consistency Algorithms for Constraint Satisfaction Problems. *Artificial Intelligence*, 25:65–74, 1985.

[54] D. Marr and E. Hildreth. *Theory of Edge Detection.* A. I. Memo 518, MIT, April 1979.

[55] J. T. McCall, J. T. Tront, F. G. Gray, R. M. Haralick, and W. M. McCormack. Parallel Computer Architectures and Problem Solving Strategies for the Consistent Labeling Problem. *IEEE Transactions on Computer*, C-34(11):973–980, November 1985.

[56] C. Mead and L. Conway. *Introduction to VLSI Systems.* Addison-Wesley, Reading, Massachusetts, 1980.

[57] B. Moayer and K.S. Fu. *Syntactic Pattern Recognition of Fingerprints.* Technical Report TR-EE-74-36, Purdue University, December 1974.

[58] Roger Mohr and Thomas C. Henderson. Arc and Path Consistency Revisited. *Artificial Intelligence*, 28(2):225–233, March 1986.

[59] U. Montanari. Networks of Constraints: Fundamental Properties and Applications to Picture Processing. *Information Sciences*, 7:95–132, 1974.

[60] MOSIS. DRA2 Chip Fabrication Report. December, 1986. Information from MOSIS.

[61] T. Pavlidis. *Syntactic Feature Extraction for Shape Recognition.* Technical Report TR-213, Princeton University, 1976.

[62] T. Pavlidis and F. Ali. A Hierarchical Syntactic Shape Analyzer. *IEEE Transactions on Pattern Analysis and Machine Intelligence*, PAMI-1:2–10, 1979.

[63] J. Pearl. *Heuristics.* Addison-Wesley, Reading, Massachusetts, 1984.

[64] J. L. Potter, editor. *The Massively Parallel Processor. Scientific Computation series*, MIT Press, Cambridge, Massachusetts, 1985.

[65] Michael J. Quinn and Narsingh Deo. Parallel Graph Algorithms. *Computing Surveys*, 16(3):319–348, September 1984.

[66] F. Robert. *Discrete Iterations.* Springer-Verlag, Berlin, 1986.

[67] A. Rosenfeld, R.A. Hummel, and S. Zucker. Scene Labeling by Relaxation Operations. *IEEE Transactions on Systems, Man and Cybernetics*, SMC-6:420–433, 1976.

[68] A. Rosenfeld and D. Milgram. Web Automata and Web Grammars. In *Machine Intelligence*, pages 3307–3324, Edinburgh University Press, Edinburgh, 1972.

[69] Sergiu Rudeanu. *Boolean Functions and Equations*. North-Holland, Dordrecht, 1974.

[70] Ashok Samal. Segmentation and Classification in a Multiprocessor Environment. December 7, 1986. Ph.D. Dissertation Proposal.

[71] Ashok Samal and Thomas C. Henderson. Parallel Consistent Labeling Algorithms. *International Journal of Parallel Programming*, 16(5):341–364, 1988.

[72] Walter J. Savitch and Michael J. Stimson. Time Bounded Random Access Machines with Parallel Processing. *Journal of the Association for Computing Machinery*, 26(1):103–118, January 1979.

[73] A.C. Shaw. A Formal Picture Description Scheme as a Basis for Picture Processing Systems. *Information and Control*, 14:9–52, 1969.

[74] F.S. Shaw. *An Introduction to Relaxation Methods*. Dover, Dover, England, 1953.

[75] Y. Shirai. Analyzing Intensity Arrays Using Knowledge about Scenes. In P.H. Winston, editor, *The Psychology of Computer Vision*, pages 93–114, McGraw-Hill, New York, 1975.

[76] K. F. Smith, T. M. Carter, and C. E. Hunt. Structured Logic Design of Integrated Circuits Using the Storage/Logic Array SLA. *IEEE Transactions on Electron Devices*, ED-29(4):765–776, April 1982.

[77] L. Snyder. Introduction to the Configurable, Highly Parallel Computer. *IEEE, Computer*, 15(1):47–56, January 1982.

[78] R.V. Southwell. *Relaxation Methods in Engineering Science. Engineering Science Series*, Oxford University Press, London, 1940.

[79] R.V. Southwell. *Relaxation Methods in Theoretical Physics. Engineering Science Series*, Oxford University Press, London, 1946.

[80] G.L. Steele and G.J. Sussman. *Constraints*. AI-MEMO MIT AI Lab. Memo 502, MIT, November 1978.

[81] G. Stockman. *A Problem Reduction Approach to the Linguistic Analysis of Waveforms*. Technical Report TR-538, University of Maryland, May 1977.

[82] K. Sugihara. Quantitative Analysis of Line Drawings of Polyhedral Scenes. In *Proceedings of the Fourth International Joint Conference on Pattern Recognition*, Kyoto, Japan, 1978.

[83] I.E. Sutherland. *SKETCHPAD: A Man-Machine Graphical Communication System*. PhD thesis, MIT, Cambridge, Massachusetts, January 1963.

[84] G. Tang and T. Huang. A Syntactic-Semantic Approach to Image Understanding and Creation. *IEEE Transactions on Pattern Analysis and Machine Intelligence*, PAMI-1:135–144, 1979.

[85] J.R. Ullmann. Discrete Optimization by Relational Constraint Satisfaction. *IEEE Transactions on Pattern Analysis and Machine Intelligence*, PAMI-4(5):544–550, September 1982.

[86] T. Vamos and Z. Vassy. Industrial Pattern Recognition Experiment - A Syntax Aided Approach. In *Proceedings of the IJCPR*, pages 445–452, Washington, DC, 1973.

[87] R.S. Varga. *Matrix Iterative Analysis. Series in Automatic Computation*, Prentice-Hall, Englewood Cliffs, New Jersey, 1962.

[88] D. Waltz. Understanding Line Drawings of Scenes with Shadows. In P.H. Winston, editor, *The Psychology of Computer Vision*, pages 19–91, McGraw-Hill, New York, 1975.

[89] W. Wang, J. Gu, and Thomas C. Henderson. A Pipelined Architecture for Parallel Image Relaxation Operations. *IEEE Transactions on Circuits and Systems*, CAS-34(11):1375–1384, November 1987.

[90] W. Wang, J. Gu, and Thomas C. Henderson. *A Systolic Array Implementation of a Discrete Relaxation Algorithm.* Technical Report UUCS-TR-86-008, The University of Utah, 1986.

[91] U. Weiser and A. Davis. A Wavefront Notation Tool for VLSI Array Design. In H.T. Kung, B. Sproull, and G. Steele, editors, *Proceedings of CMU Conference on VLSI System and Computations*, pages 226–234, Computer Science Press, Rockville, Maryland, 1981.

[92] P. Winston. *Artificial Intelligence.* Addison-Wesley, Reading, Massachusetts, 1984.

# Index